Searching for ORDER IN the COMPLEXITY of Evolving Worlds

AUTHOR ACKNOWLEDGMENTS

I would like to express my gratitude to Sienna Latham, Zato Hebbert, Rachel Fudge, David Krakauer, and Tasos Psychogyiopoulos for their invaluable contributions to the online and print versions of the book. I would also like to thank Jos Verbrugge, Robert Goldstone, Mirta Galesic, Henrik Olsson, and John Miller for their valuable reviews of the preliminary version of the book.

PRESS ACKNOWLEDGMENTS

The SFI Press would not exist without the support of William H. Miller and the Miller Omega Program.

[COMPLEX-SYSTEMS RESEARCH IN PSYCHOLOC

HAN L. J. VAN DER MAAS

SFI PRESS
THE SANTA FE INSTITUTE PRESS

1399 Hyde Park Road
Santa Fe, New Mexico 87501

Complex-Systems Research in Psychology
ISBN (PAPERBACK): 978-1-947864-63-4
Library of Congress Control Number: 2024945913

The SFI Press is supported by
the Miller Omega Program.

The free online version can be viewed at:
https://santafeinstitute.github.io/ComplexPsych/

We have an increasing need to model ourselves.

J. DOYNE FARMER (2[illegible])

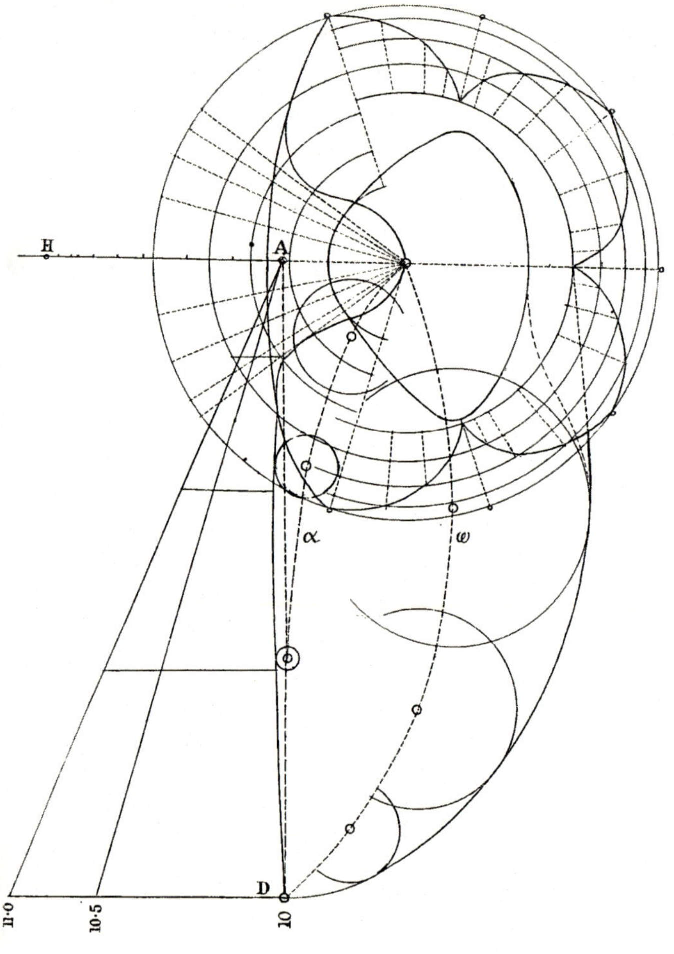
H
A
D
α
ω
11·0
10·5
10

CONTENTS

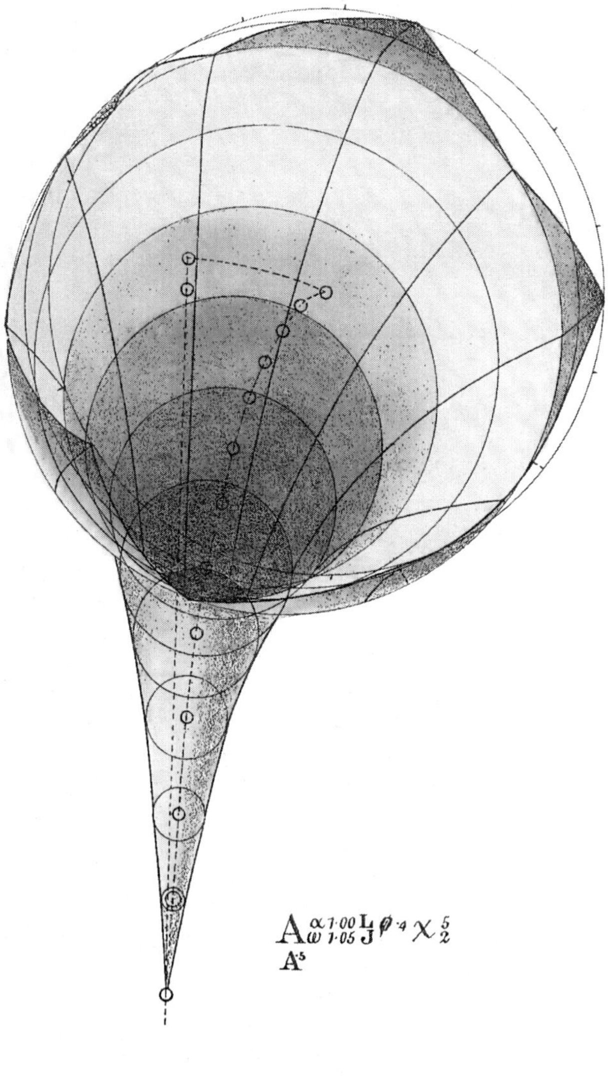

$A^{\alpha\,1{\cdot}00}_{\omega\,1{\cdot}05}\,{}^{L}_{J}\,\phi^{\cdot 4}\,\chi\,\frac{5}{2}$

$A^{\cdot 5}$

PROLOGUE

This book is intended for psychologists and social scientists interested in modeling psychological processes using the tools of complex-systems research.

The book has three primary objectives. The first is to provide a comprehensive overview of complex-systems research, with a particular emphasis on its applications in psychology and the social sciences. The second is to provide skills for complex-systems research. Lastly, it strives to foster critical thinking regarding the potential applications of complex systems in psychology.

RESOURCES
The electronic version of this book, including code blocks and links, can be viewed and accessed at: *santafeinstitute.github.io/ComplexPsych/*

For many decades, with roots dating back to the nineteenth century, scientists have been studying a wide variety of complex systems. Well-known examples include lasers, tornadoes, chemical oscillations, ant nests and flocks of birds. Scientists have built mathematical and computational models of these complex systems and developed techniques to study them.

Applying these techniques requires a great deal of mathematical and technical knowledge, as well as deep understanding of the nature of the system. You don't just create a mathematical model off the top of your head. In addition, testing such models requires extensive and reliable quantitative data. Applying complex-systems theory to the behavioral and social sciences is therefore not straightforward. Theories are often verbal, and quantitative measurement in these sciences is a longstanding issue. While there has been some reasonable progress over the past 150 years, it is fair to say that the behavioral and social sciences are less mature than the "hard" sciences.

Despite these challenges, applying complex-systems theory to the behavioral sciences is imperative. Whether we consider humans in isolation, the billions of interacting neurons in the brain, or the social networks in which we find ourselves, complexity is everywhere. We, with our complex brains embedded in various hierarchies of social systems, are the ultimate complex systems.

I believe that we can only succeed in exploring the psychological system by understanding its complexity. We need to apply the tools of complexity science to psychology, which is in desperate need of breakthroughs. After all, the modern world revolves around human beings who, through language and thought, have created an unimaginably complex world. The greatest danger now are humans ourselves, and progress in the field of psychology is necessary and urgent.

This book requires study. New theoretical concepts are illustrated with simulations and examples. Running the simulations, studying the examples, and solving the exercises will contribute to a deeper understanding of the material. I have used the book's content in a master's course for research-minded students in psychology. Readers should have some background in psychology and its research methods.

I assume only pre-university knowledge of mathematics. An important prerequisite is a basic knowledge of the programming language R. The book uses R for the simulations and exercises. There are many online resources for learning the basics of R. In addition to R, we will use NetLogo, but no prior knowledge of NetLogo is expected. NetLogo is a dedicated programming language for simulating complex natural and social phenomena. It is freely available for all major computer platforms.

This book has been made available as open access courtesy of the Santa Fe Institute Press, a non-profit publisher. Consistent with this approach, I often cite open sources such as Wikipedia rather than proprietary ones. It should be noted that open sources such as Wikipedia have the potential for content changes, although this is unlikely in the contexts I've referenced. I also use only open-source software for the examples and exercises.

It should also say that this book is more a book for psychologists who have very limited knowledge of complex-systems research than the other way around. Experts in complex systems who wonder how it can be applied in psychology may have to wait for another text.

I have written this book based on thirty-five years of scientific work in collaboration with fantastic colleagues and coauthors of many papers. I'm part of the ecosystem of the psychology department, especially the wonderful methods section, of the University of Amsterdam. Also important is the Institute for Advanced Study in Amsterdam, which has complex-systems research as a central theme. In recent years, I've also been an external faculty member at the renowned Santa Fe Institute in New Mexico. I am indebted to all of them and to many other colleagues around the world. In my citations, I've made an effort to acknowledge the extensive contributions to this vast field. Nevertheless, I recognize that there may be omissions, for which I apologize.

Han L. J. van der Maas, Amsterdam 2024

INTRODUCTION

1.1 What are Complex Systems?

Some things in life are simple. When you push a block, it moves. Pushing harder makes it move faster, and stopping the push stops its movement due to friction. When you open the tap, water starts to flow. If you open it farther, it flows faster, and if you close it, it stops. Cause-and-effect relationships like this are clear and roughly linear. Such relationships are rare in psychology. Let's take fear as an example. A fear stimulus—for example, a barking dog—can lead to fear and flight, but also to anger and attack. Whether the stimulus is perceived as fearful might depend on subtle differences in context. It has also been debated whether the flight response precedes the feeling of fear or vice versa. Fear could also suddenly change into a panic attack.

RESOURCES
The electronic version of this book, including code blocks and links, can be viewed and accessed at: *santafeinstitute.github.io/ComplexPsych/*

In psychology, cause-and-effect relationships are rarely simple, and effects are often nonlinear.

These difficulties are not unique to psychology. Many systems studied in physics, chemistry, and biology show such complex behavior (Weaver 1948; Krakauer 2024). They are complex systems. Although there is no full consensus on the definition of a complex system (Ladyman, Lambert, and Wiesner 2013; Heylighen 2009), I believe the core aspects can be summarized as follows.

Complex systems are made up of many smaller interacting subsystems, such as atoms, molecules, cells, neurons, and even entire organisms. I like the term *subsystems* because the lower-level elements can themselves be complex systems.[1] The interactions between subsystems can be of different kinds, but they are generally local, fast, and nonlinear. These interactions result in emergent behavior. The emergent processes usually operate on a slower time scale. A typical example, which will be discussed in more detail later, is the traffic jam. Cars react mainly to cars in their vicinity, which can lead to global patterns of congestion. An-

Complex systems exhibit emergent behavior, meaning that the interactions between the units of the system result in global patterns or properties that do not occur in the units themselves.

[1] As Simon (1962) noted, atoms were once considered elementary particles, whereas in modern nuclear physics they are themselves complex systems.

other example is magnetism which is not present in any of the atoms of the magnet.

Self-organization is a process in which some form of overall order or coordination develops from the local interactions between the parts of an initially disordered system.

In general, these patterns emerge through self-organization. An example is ants building an ant nest: no one ant oversees or directs this process. Rather, it emerges from the local interactions between many ants.[2] I will explain this in more detail in chapter 5, but in a completely closed system, self-organization would not be possible. Self-organization takes energy. The emergent patterns in a complex system may be stable for some time, but often change suddenly. The study of phase transitions or tipping points is therefore central to the study of complex systems. They may also exhibit chaotic behavior, implying that they can be fundamentally unpredictable, the weather being a notorious example.

Complex systems are open systems, meaning that they use energy that they have absorbed for the environment.

Let's look at one famous case, the flocking of birds (fig. 1.1). Flocks of birds move in a beautiful choreography as they glide through the air, their formations shifting and morphing as they twist and turn across the sky. Flocks are well understood and easy to simulate, as we will see in chapter 5. Flocks fulfill all the criteria of a complex system (see Parisi 2023 for an extended analysis). They are open systems, as birds use energy to fly. Each bird responds only to birds in its local neighborhood. They follow roughly three rules: they try to fly in the same direction as their neighbors, stay close to their neighbors, and avoid collisions. These are fast and local interactions. If you watch some videos of flocks of birds on the internet, you will see globally organized behavior. This is a prime example of self-organization. There is evidently no one bird in control ordering other birds to change direction. What you can also see in these movies is that stable patterns, say an oval shape, can suddenly change. The birds may change direction or split up. Such bifurcations or catastrophes (to be explained in ch. 3) are typical of complex systems.

This globally organized behavior of a flock is a form of spontaneous order.

[2] It is also argued that emergence is a consequence of symmetry breaking (Krakauer 2023). Symmetry breaking occurs when a system transitions from a symmetric state to an asymmetric state, resulting in the emergence of distinct properties or behaviors. An example of symmetry breaking can be observed in the formation of snowflakes. Initially, ice crystals have a symmetrical hexagonal shape due to the underlying molecular structure of water. However, as the crystal continues to grow, environmental factors such as temperature and humidity influence its growth pattern. Minute variations in these factors lead to the breaking of initial symmetry and the formation of diverse and beautiful snowflake structures.

Figure 1.1. A flock of birds is an example of a complex system in which global patterns emerge naturally from simple rules and local interactions.

You can also see that the behavior of these flocks is rather unpredictable. Flocks, and swarms in general, are well understood and can be easily simulated on a computer, but this does not mean that we can always predict these systems, an issue that will be discussed further in chapter 2.

Similar examples can be found in all natural sciences. Tornadoes, for example, are made up of air molecules that also interact locally. Tornadoes are unpredictable, self-organizing, global weather phenomena. A famous chemical example is the Belousov–Zhabotinsky reaction, a chemical oscillator (Kuramoto 1984). This reaction involves a mixture of chemicals. As the reaction proceeds, the solution exhibits strikingly colorful oscillations between clear and opaque or between different colors, depending on the specific reactants used. We will see many more examples in later chapters.

The prime example from psychology is the brain. About a hundred billion neurons interact with thousands of other neurons in their neighborhood. Compared to computers, brains are extremely energy-efficient, but like all open systems, they do consume energy (the equivalent of a light bulb according to Attwell and Iadecola 2002). The letters you are reading activate retinal neurons that initiate a cascade of electrical waves across billions

Fast local interactions somehow form global waves of electrical activity that make up thought processes and even consciousness.

of neurons that somehow create your understanding of this text (Roberts et al. 2019; Schöner 2020). How is this possible? For me, this is the most fascinating scientific question of all time. It's the main reason why I'm a psychologist and not a physicist. I view the brain as the ultimate complex system.

1.2 Emergentism

What is the relation between complexity and reductionism? According to reductionism, complex phenomena can be explained by reducing them to the interactions of their individual parts or components. This raises two questions, one related to weak emergence and one related to strong emergence.

The first question is, why it is possible to conduct scientific research in fields other than physics, given that, ultimately, chemistry, biology, and even psychology are fundamentally concerned with interactions among elementary particles. Should we not first finish the study of physics before starting to think about complex molecules, cells, neurons, or higher-order human cognition?

Science is possible at many different levels of description without fully understanding the lower levels.

Philip Anderson's renowned paper "More Is Different" convincingly argues that the answer to this question is a resounding *no* (Anderson 1972). There is much to be said for reductionism, but somehow the laws of quantum mechanics are irrelevant when studying interactions between neurons or people. I don't think that emergence in complex systems is inconsistent with a reductionist view of science (Bechtel and Abrahamsen 2005). One could say that complex-systems theory explains why emergent phenomena such as atoms or neurons can be used as lower-order entities at even higher levels of description to explain new higher-order phenomena, without being a dualist.

"In the face of complexity, an in-principle reductionist may be at the same time a pragmatic holist" (Simon 1962).

This fundamental principle of emergence is what allows disciplines like psychology to exist as distinct and independent fields of science (Fodor 1974). The concept of level is central to Herbert Simon's architecture of complexity, in which each subsystem is itself a complex structure made up of smaller parts, and this pattern is repeated at multiple levels.[3] According to Simon, these nested structures are ubiquitous in the natural world and

[3] See figure 6.4 for a visual illustration of this idea.

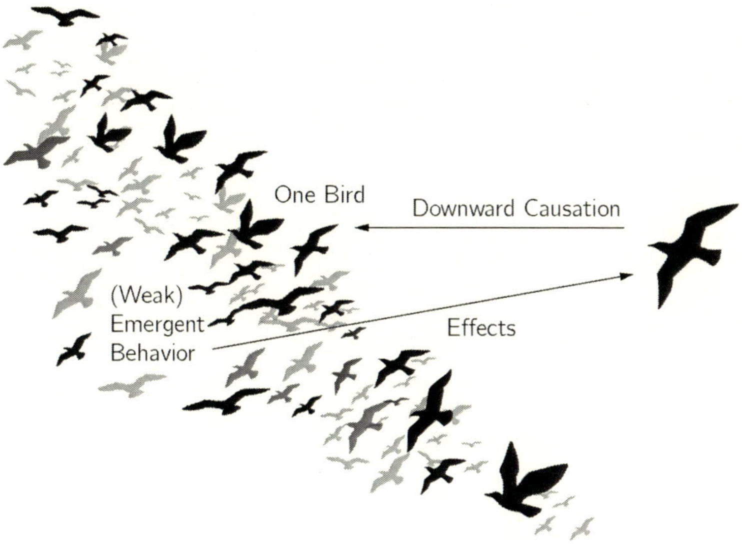

Figure 1.2. An illustration of downward or circular causation in flocks due to a predator responding to the emerging patterns of the flock and subsequently influencing the flight of individual birds.

in human-made systems because they are robust and adaptable. For an in-depth discussion of this level concept I refer to Wimsatt (1994).

The second, more controversial, question is whether emergent phenomena have an independent causal role (strong emergence) or mainly have descriptive value (weak emergence). Strong emergence is often associated with downward causation (Chalmers 2006; Flack 2017; Kim 2006).

Downward or circular causation is the idea that higher-level entities or properties can influence the behavior of lower-level entities or properties.

I like to link this to the flocking example. Flocks of birds are emergent phenomena that do not determine the behavior of the individual birds. The birds only follow the local rules. Flocking is an example of weak emergence. However, when predators enter the scene, things change. Predators get confused by flocks of prey, not by the behavior of individual birds. So the flock has some causal power. Moreover, the birds react to the predator's movements. This could be seen as an example of downward causation and thus strong emergence (fig. 1.2). Recent work attempts to quantify such causal emergence effects (see, e.g., Hoel, Albantakis, and Tononi 2013).

Both weak and strong emergence are essential to understanding psychological phenomena.

Our minds, encompassing conscious thought, self-awareness, reasoning abilities, natural language comprehension, emotions, and attitudes, are not mere artifacts and cannot be simply reduced to intricate patterns of neural activity. In my view, these mental constructs, such as consciousness, possess their own causal influence, and this is one of the reasons that psychology stands as a scientific discipline in its own right.

Of course, the relationship between the mind and the brain is one of the most debated topics in psychology. The neuro-reductionist view is popular in the field of psychology, when it concerns the explanation of both higher cognition (Schwartz et al. 2016) and psychological disorders (for a critical review, see Borsboom, Cramer, and Kalis 2019).

1.3 The Field of Psychology

The study of complex systems in the natural sciences is highly technical. I like to think of the field of complex systems as a toolbox of empirical paradigms and mathematical models and techniques (Grauwin et al. 2012). Models are often formulated in the form of difference or differential equations and subjected to, for example, bifurcation analysis. These are mathematical ways of describing the behavior of complex systems. Additionally, advanced numerical analysis, commonly in the form of computer simulation, is a standard approach. However, educational programs in psychology do not usually include courses in algebra, calculus and programming. Many psychologists lack the basic knowledge and skills to apply the toolbox of complex-systems theory, as these are not ordinarily part of the psychology curriculum. Complex-systems research simply seems too complex for psychologists and social scientists. One goal of this book is to provide psychologists with a first introduction to this technical toolbox.

Unfortunately, there are additional complications in applying these tools to our field. First, our subjects are much more complex than flocks of birds or tornadoes and they display astonishing behavior. They can do science! They can also walk out of the lab because they find the experiment boring. This does not happen with lasers. Second, we have to deal with the ethical constraints of experimenting on our subjects. We cannot take them apart, a very successful approach in the natural sciences. Finally, there is the measurement issue (Lumsden 1976; Michell 1999).

We tend to forget how incredibly precise the natural sciences, especially physics, are. In 1985, Richard Feynman famously claimed that the accuracy of calculating the size of the magnetic moment of the electron was equivalent to measuring the distance from Los Angeles to New York, a distance of over 3,000 miles, to the width of a human hair. I find that shocking. Less famously, I would argue that psychologists have not yet "discovered" continents and have no idea where New York is. Our instruments generally fail to meet elementary requirements of reliability and validity, we are plagued by replication failures, and our theories are often imprecise (Eronen and Bringmann 2021).

Navigating the behavioral and social sciences and knowing which data to trust and which empirical phenomena to model is an art in itself.

This is all unfortunate because not only our brains but every subject in our field seem to have the characteristics of a complex system. Social systems are complex systems made up of individuals interacting to produce emergent phenomena such as cultures and economic systems. The human brain, arguably the most complex system we know, is embedded in different hierarchies of very complex social systems such as families, education, economies, and cultures. We need the toolbox!

Despite all these problems, I'm not pessimistic. I believe that tangible progress in the behavioral and social sciences is possible. It is not that these sciences are completely unsuccessful. We know a lot about people's attitudes, addictive behavior, cognition, and the social systems in which they interact. We study these, with some success, using advanced experimental designs, and we have developed (mainly) verbal theories about almost everything.

We also have no choice; we must make progress. Personally, I feel a strong tension between our struggle to elevate the behavioral and social sciences as a science, on the one hand, and the enormous expectations of society to deliver, on the other. Our most pressing global problems—climate change, overpopulation, war and violence, poverty, inequality, infectious diseases, addiction, to name but a few—are unsolvable without breakthroughs in the behavioral and social sciences.

J. Doyne Farmer: "We have an increasing need to model ourselves" (Thurner 2016).

The realization that the human mind in its social context is an amazingly complex system also offers opportunities. Despite their obvious differences, complex systems show remarkable similarities. A predecessor of complex-systems theory, general-systems theory (Bertalanffy 1969), explicitly assumed that all systems share important characteristics. This is the primary reason for providing numerous modeling examples in this book that originate from disciplines beyond psychology.

Certain mechanisms and phenomena seem to operate and to occur in similar ways at all possible levels of description (Simon 1962).

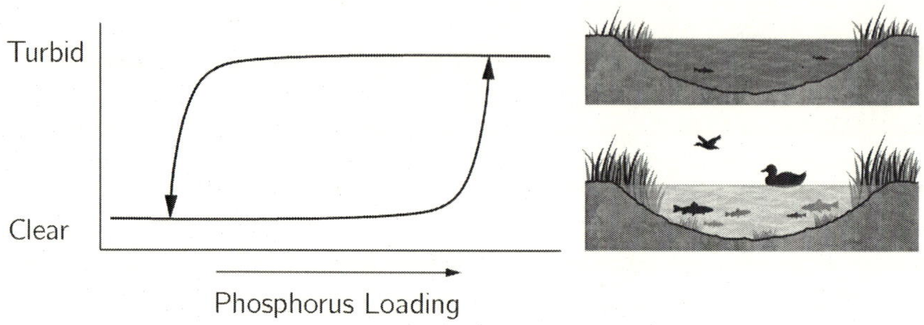

Figure 1.3. The transitions between clear and turbid states of shallow lakes do not occur at the same phosphorus load. This delay in jumps is called hysteresis. Hysteresis explains why transitions are often difficult to reverse. This concept is discussed in detail in chapter 3.

An inspiring example for me comes from the study of shallow lakes (Scheffer 2004). Shallow lakes tend to be either in a "healthy" state, with clear water and a diverse population of fish and plants, or in an "unhealthy" turbid state. I like to compare these complex lake systems in the turbid state to a person suffering from depression. This turbid state usually occurs suddenly. There is a critical phosphorus load at which the system turns over from being healthy to complete dominance by algae and bream. Typical of this type of transition is the hysteresis effect (fig. 1.3). This means that the turning point from clear to turbid and from turbid to clear does not occur at the same phosphorus load. The turning point to clear water only occurs at much lower phosphorus loads. These tipping points may be so far apart that reducing the cause, the phosphorus load, is not a viable option. Of course, all sorts of interventions have been studied, such as supplemental oxygen, chemicals, sunscreens, and stocking predatory fish. These interventions have not been very successful, or only successful in specific cases. The fact that they had some level of success brings to mind the partial effectiveness of clinical interventions, such as those used in the treatment of major depressive disorder.

A breakthrough occurred in the 1980s. Removing all the fish proved to be a very effective intervention. The ecologists caught almost all the fish with nets during the winter. In the spring, a new, healthy equilibrium emerged, characterized by aquatic plants, other fish species, and clear water. This new state is often stable for long periods of time. Remarkably, the analysis of the cause, the phosphorus load, was not part of the solution: although the increasing phosphorus load is the primary cause of the transition toward a turbid state, decreasing the phosphorus load does not cause the system to transition back into the clear state. What this means for our thinking about depressive disorders will be discussed in chapters 5 and 6.

The dogma of intervention, that the cause of the problem is the key to the solution, does not necessarily apply to complex systems.

I will discuss three specific reasons to be somewhat optimistic, based on three key observations about complex systems. The first key observation has to do with simplification, the second with the tendency of complex systems to be characterized by a limited number of stable states, and the third is that all complex systems seem to be describable as some kind of network. Simplification is perhaps the most important one.

1.4 The Art of Simplification

Einstein supposedly said that everything should be made as simple as possible, but not one bit simpler.

A fascinating and instructive example is the traffic jam, which is made up of many people, with their amazingly complex brains, in modern cars full of advanced technology. Where to start modeling such a complex phenomenon? The answer is astonishing. It seems that we can reduce people in cars to simple blocks in a lane, speeding up when there is space in front of the artificial car and braking when they get too close to the car in front. All lower levels of modeling are ignored. This is even simpler than a flock of birds.

It is not difficult to set up a computer simulation for this case. I recommend that you spend some time playing around with an example.[4] It does not take long to see that traffic jams can easily form and have an unexpected property: while cars move forward, traffic jams move backward! Another interesting observation is that variance in speed causes congestion. But the variance is not in any of the cars. Variance and congestion are properties at a higher level of description. With this simulation, you can study different types of traffic situations and interventions. This type of simulation proves very useful for designing and optimizing highways and roads (Barceló 2010; Treiber, Hennecke, and Helbing 2000). There are actually different ways to model traffic jams. Jusup et al. (2022) distinguish between fluid-dynamical, kinetic, car-following, coupled-map lattice, and cellular automata models. They all reproduce many phenomena of real traffic jams.

[4] A nice example is https://www.traffic-simulation.de/.

In complex systems, the qualitative properties of large-scale phenomena do not depend on microscopic details.

Only higher-level properties are relevant to global behavior. A large part of the art of science is finding the right level of simplification. Suppose we are studying smoking. Do we model the effects of nicotine on blood vessels, how the hand with the cigarette moves from the mouth to the ashtray, or the number of cigarettes smoked per day? Do we include the effects of marketing and the smoker's social network?

What is relevant and what can be ignored? It can be challenging to provide a definitive answer for specific cases. Nevertheless, in general, it can be said that there is a limit to the lower levels that must be considered. When examining traffic jams, it is necessary to incorporate certain characteristics of individuals and vehicles, but delving deeper into topics like neuronal firing, DNA replication, or the intricate workings of car batteries becomes irrelevant. At that level of modeling, there is no relevant information that would alter the explanation of a traffic jam.

The traffic example shows that extreme simplification is sometimes possible and necessary. But finding the right level of simplification is not a simple task at all. In Borsboom et al. (2021) we propose a *theory-construction methodology* (TCM) consisting of five steps:

1. Identify the empirical phenomena that become the target of explanation.
2. Formulate a set of theoretical principles that putatively explain these phenomena.
3. Use this set or prototheory to construct a formal model, a set of model equations that encode the explanatory principles.
4. Analyze the explanatory adequacy of the model, that is, whether it actually reproduces the phenomena identified in step 1.
5. Determine whether the explanatory principles are sufficiently parsimonious and substantively plausible.

The article explains these steps in detail and provides an example, the mutualism model of general intelligence, which is explained in chapter 6 of this book.

I will add a few comments to this list of steps. First of all, step 1 is key. It is crucial to be precise about defining the phenomena to be explained. Phenomena are not the same as data. Data are particular empirical patterns (a concrete dataset), whereas phenomena are general empirical patterns, stable and general features of the world (Haig 2014). As noted above, in the behavioral and social sciences, it is not always clear which data patterns can be trusted. In the last ten years, the replication crisis has led to a revolution in psychological methods, but many results are collected using potentially biased methods. One problem is publication bias: negative results are still harder to publish than results that support hypotheses. In other cases, the results of different studies contradict each other, and meta-analyses show weak effects at best.

Draw[illegible] list of the most i[illegible] phenomena on a topi[illegible] s depressio[illegible] ng, or discriminatio[illegible] en a challenge.

The second observation is that taking these steps is not a linear process. Often, when developing a model, you realize that some important information is missing from the list of phenomena. For example, you might be modeling addiction, but suddenly you need information about the combination of addictive substances that people use. And such simple questions are often impossible to answer. You can spend days searching the literature for information that you expect to be readily available, only to find that many basic things are simply unknown.

A third observation is that formal modeling is mostly a matter of analogical reasoning. You have to study many examples of complex-systems models to understand how to construct such a model. Indeed, in my own work I often use established models developed in physics and biology as a base model. We will see many examples of this later.

Fourth, good models do not build in phenomena but explain them from basic principles. By building in, I mean that a phenomenon should not be an assumption of the model. An example would be a model that says that variance in the speed of cars causes traffic jams. Such a model may explain other things, but not the role of speed variance, because that effect is part of the assumptions. Models that make such assumptions are called phenomenological models. We will see examples of phenomenological models of complex systems, as well as explanatory models, where the latter are based on fundamental principles.

Building real explanatory models in our fields is extremely difficult.

Fifth, it is my conviction that a metaphorical use of the complex-systems approach should be avoided by using concrete formal models (Dongen et al. 2024). It is crucial to strive for the highest level of scientific rigor. There are no special, more lenient, methodological rules for complex-systems research (van der Maas 1995).

1.5 A Limited Number of Equilibria

The first key observation was that complex systems can be simplified. The second is that complex systems tend to be characterized by a limited number of equilibria. An important example is water. Water normally exists in either a solid, liquid, or gaseous state (leaving aside the plasma state). These are stable states over wide ranges of temperature and pressure.

A biological example is the life stages of a butterfly (egg, caterpillar, chrysalis, and butterfly). Except for brief periods of transition, these insects are in one of these four relatively stable states. Another example is the horse, which is either standing still, walking, trotting, or galloping. I am convinced that we must always start by identifying the equilibria of a complex system. This also applies to psychological and social science applications. A bipolar disorder seems to be characterized by two stable states (depressive and manic). In the case of addiction, we may think of a three states: non-use, recreational use, and heavy use (Epskamp et al. 2022). Similarly, we could identify the stages in falling in love, in understanding of calculus, in sleeping, and in radicalization.

There is an ongoing discussion about the number of stages, even for something like sleep (Boostani, Karimzadeh, and Nami 2017; de Mooij et al. 2020).

Identifying discrete stages turns out to be more difficult than it first appears. It is often possible to come up with more substages. For instance, in the case of horse movement, people tend to further subdivide trot into three forms (working, medium, and collected). Subdivisions are also made in the case of heavy alcohol consumption (Leggio et al. 2009). It is possible to use objective statistical methods to support such classifications using modern machine-learning techniques (automatic clustering) as well as more traditional means (finite mixture models, latent class analysis). I will say more about this in section 3.5.

A further complication is that equilibria come in different forms. The simplest form consists of fixed points or point attractors, an example being a ball lying in a valley. Under undisturbed conditions, the ball could also be resting on top of a hill, which is an unstable equilibrium. An equilibrium could also be a limit cycle or oscillator. For example, two pendulums could swing in phase or out of phase. It gets even weirder when we consider strange attractors, which often take the form of fractals. This will be explained in more detail in the next two chapters.

Finally, it has also been argued that many complex systems, especially living systems, never reach equilibrium because they are constantly perturbed (Groot and Mazur 2013). But at least some complex psychological systems are clearly stable over the long term. Unfortunately, this is true of many psychological disorders. In contrast, my understanding of the world, psychological science, and complex-systems research is better described as a continuously perturbed non-equilibrium system with just enough stability to write this book (once).

I see this distinction between equilibrium and non-equilibrium complex systems as gradual.

I would claim that many psychological complex systems tend to be in one attractor state most of the time, but they occasionally change states. If certain control parameters slowly change their values, the current equilibrium can become unstable and a transition to another equilibrium can occur. This is what happens when we lower the temperature of water to below zero. The family of transition models is described by bifurcation theory. This is explained in chapter 3, where we focus on a very important transition model, the cusp catastrophe, and in chapter 4, which considers dynamical systems models.

Transitions can occur in many ways, also depending on the types of equilibria involved.

1.6 Networks are Everywhere

The third key observation of great relevance to the attempt to use complex-systems modeling in psychology is that complex systems are networks, as they consist of interacting subelements. For me, the network is the most interdisciplinary research topic in modern science. Magnets, ecosystems, the brain, the internet, and social networks are prime examples.

Network science is a huge area of research with many fundamental insights and an important tool in modern psychological science.

Two applications in psychology are well known: the first is the study of neural networks, which started seventy years ago and has become the main foundation of the artificial intelligence (AI) revolution of the last ten years. In chapter 5 I will discuss neural networks. The second is social networks, the simplest example being dyadic interactions. Social media such as Facebook are infamous examples. Key ideas relate to concepts such as weak and strong ties, central hubs and homophily, which are discussed in chapters 6 and 7. The analysis of social-network data is an exciting area of research (Scott 2011). It focuses on understanding how social entities are connected and how these connections influence various outcomes and behaviors. Connections between nodes (e.g., individuals, organizations, communities) can be based on different dimensions, such as friendship, communication, collaboration, information flow, or any other form of social interaction. These interactions may also change over time, which is studied in social-network dynamics (Snijders 2001).[5]

Chapter 6 focuses on a novel use of networks, which I call *network psychology*. This is a level of description between neural networks and social networks. It involves modeling intelligence, attitudes, and psychological disorders at the individual level. Intelligence, for example, is modeled as an ecosystem of cooperating cognitive functions. This is radically different from the standard view that general intelligence is due to g, a single underlying source. In the mutualism model of general intelligence (van der Maas et al. 2006), the observed positive correlations between scores on subtests of IQ test batteries are due to cumulative reciprocal developmental interactions between cognitive subsystems such as working memory, spatial cognition, and language.

Similarly, depression can be thought of as a network of mutually reinforcing symptoms.

As another example, sleep problems, a symptom of depression, can lead to increased fatigue and difficulty concentrating, which in turn can affect a person's ability to manage daily tasks and engage in social activities. This can then lead back to poorer sleep quality, creating a cyclical pattern in which each symptom reinforces the others. This new view of mental disorders originated in our research group and is now very popular (Robinaugh et al. 2020). One reason for this is that many statistical techniques have been developed to investigate this network approach.

The latest line of this research is the integrated study of psychological and social networks (van der Maas, Dalege, and Waldorp 2020). Chapter 7 deals with models in which psychological network models of attitudes are nested within social networks of opinion change. This model provides a new explanation of polarization.

In the process of polarization, nonlinear intrapersonal and interpersonal dynamics interact.

1.7 Methods for Investigating Complex Systems

Complex systems are studied in various ways across different disciplines. We use computer simulations to examine the emergence in complex system models, analyze their unpredictable behavior, categorize the types of tipping points involved, derive equations that describe the overall behavior of

[5] Statnet.org provides an overview of R packages for social-network analysis.

complex systems, collect and analyze time-series data, or experimentally disrupt the system to test its resilience. Following Sayama (2015), I categorize the models and methods into two groups: those for systems with a small number of variables and those for systems with many variables.

The first category is referred to as *nonlinear dynamical system theory*, which encompasses chaos theory and catastrophe or bifurcation theory. The second category includes tools for studying multi-element systems, such as agent-based modeling and network theory. One might assume that the first category is irrelevant to complex systems, which by definition have many variables. However, it has been found that the global behavior of complex systems can often be described with a small number of variables, often just one, that behave in a highly nonlinear manner.

I consider nonlinear dynamical system theory an essential part of complex-systems research.

This categorization is reflected in the book's structure. The next three chapters are devoted to systems with a small number of variables: chapter 2 discusses chaos theory, chapter 3 addresses sudden transitions as studied in catastrophe and bifurcation theory, and chapter 4 provides an introduction to modeling dynamical systems.

In the second part of the book, we shift our focus to tools for studying systems with many variables, particularly agent-based modeling of self-organization in Chapter 5, network modeling in chapter 6, and the application of both to psychosocial systems in chapter 7.

1.8 Other Work and Sources

The complex-systems approach has often been introduced as the next new thing, but those days are gone. Even in psychology it can no longer be considered a new approach. Many different research groups have used the toolbox of complex-systems research in all areas of psychology. This book will give many examples. One could even argue that a lot of work has been done that could be considered complex-systems research but has not been published under that heading. For example, most neural-network models of psychological processes are complex-systems models because they investigate emergent computational properties of the interaction of neural units. This is also true of much work in mathematical psychology, for example when differential equations are used to study dynamical systems. Older work in complex-systems research has often been published with reference to nonlinear dynamical systems. Other related approaches are computational social science and agent-based modeling.

Today, there are many interdisciplinary centers or hubs for complexity research. The Santa Fe Institute in Santa Fe, New Mexico, is the pioneer of complexity science. Its summer schools are highly recommended. Other examples are the Complexity Science Hub in Vienna, Austria, and the Centre for Complexity Science at the University of Warwick, in England. In my own country, the Netherlands, we have at least four of these centers. I'm a principal investigator at the Institute for Advanced Study in Amsterdam and an external faculty member at the Santa Fe Institute.

It is impossible to give a balanced review of all past and ongoing work on complex systems. I'm naturally somewhat biased toward our own work and contributions, but I do my best to point out relevant work. As a general resource to complex-systems research with a bit less technical approach, I recommend the book of Mitchell (2009); for a bit more mathematical approach I recommend the books of Serra and Zanarini (1990), Sayama (2015), and Thurner, Klimek, and Hanel (2018). Overviews of work in psychology are provided by Guastello, Koopmans, and Pincus (2008) and Port and Gelder (1995). Other great books are written by Heath (2000) and Kelso (1995).

1.9 Exercises

1. Visit https://www.traffic-simulation.de/. In what direction do traffic jams move? For roundabouts: What is a bad priority rule? Do traffic jams appear and disappear for the same values of critical parameters? Take for instance the ring road and vary Politeness. (*)

2. Give your own example of a psychological process or theory where different stable stages or states are distinguished. (*)

3. Could consciousness be seen as a process of downward causation? Explain your answer. (**)

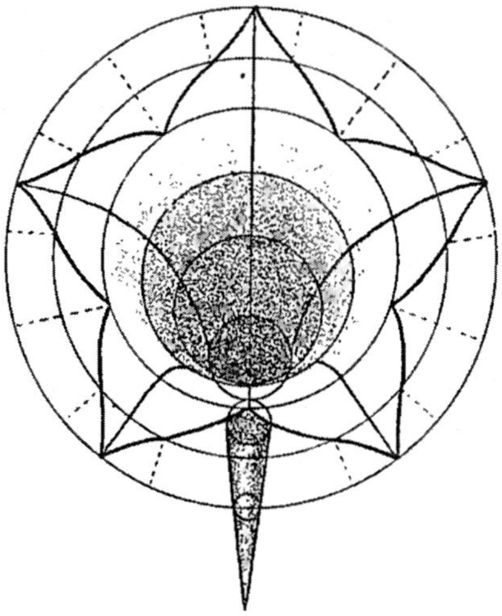

Ω FH $\phi^{\cdot 75}$ χ^{5}
A1

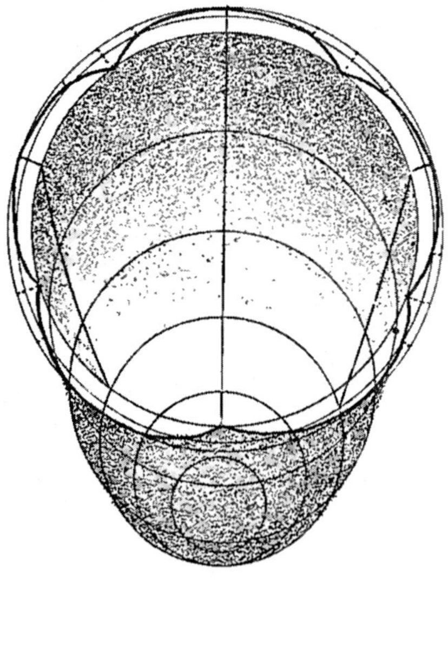

2

CHAOS AND UNPREDICTABILITY

2.1 Introduction

Suppose we have immense amounts of genetic, biological, and psychological data on millions of participants and knowledge of all relevant environmental factors. Suppose also that these huge amounts of data are of fantastic quality. Using state-of-the-art machine-learning models and powerful computing resources, we could build advanced statistical models that include main and higher-order interaction effects of all variables, even incorporating nonlinear transformations. Even then, prediction may not be possible. Why? Because of a phenomenon called *deterministic chaos*.

RESOURCES
The electronic version of this book, including code blocks and links, can be viewed and accessed at: *santafeinstitute.github.io/ComplexPsych/*

Deterministic chaos refers to the behavior of complex systems that are highly sensitive to initial conditions, leading to unpredictable and seemingly random results despite being governed by deterministic laws.

Chaos is one of the most spectacular phenomena in complex systems, and as psychologists we should know the basic results of chaos theory. It is also great fun to learn about chaos and it allows me to introduce many key concepts that we need in later chapters.

In my opinion, the direct applicability of chaos theory to psychology and social science is somewhat limited. For a long time, researchers have tried to show chaos in time series of psychophysiological measures, but this seems to be difficult. I will briefly review this work at the end of the chapter. The relevance of chaos theory may lie not in its application but in its fundamental implication for prediction. What chaos theory basically shows is that even in the best of circumstances, where we have very accurate models and data, long-term prediction might be impossible.

This is known as the butterfly effect (E. N. Lorenz 1963): a butterfly flaps its wings in India, and that tiny change in air pressure could eventually cause a tornado in Iowa.

2.2 The Population Growth of Rabbits

Chaos theory consists of many deep mathematical results, but understanding the basics of chaos is not so hard. Below I will explain chaos in difference equations at a very basic level of mathematics and programming. The elementary example is the famous logistic map, usually introduced as a model of population growth, for instance, of rabbits. Suppose we have rabbits on an island, and they start to multiply. What would the mathematical model be for such a process?

Population growth is a typical example of a dynamical system, as it is a model of change.

In general, in a dynamical system, the change or growth of a variable (say X) depends on the current state and some parameters. Time plays a very special role. We can use discrete or continuous time steps. In the first case, which is the focus of this chapter, we use difference equations; in the second case, we use differential equations. In the logistic map, time is discrete (population growth takes place in generations). The simplest dynamical model for the population growth of rabbits is

$$X_{t+1} = rX_t. \tag{2.1}$$

This says that the new value of X is determined by the previous value of X, multiplied by r. In this equation r is the growth rate. We can simulate this model by choosing a value for r, $r = 2$, for instance. We also need an initial value, say $X_0 = 1$. If this is completely new to you, enter some values repeatedly. You will see exponential growth ($X_1 = 2$, $X_2 = 4, X_3 = 8, X_4 = 16$, *etc.*). In R we can simulate this using a for loop (the result is shown in fig. 2.1).

```
n <- 15
r <- 2
x <- rep(0,n)
x[1] <- 2 # initial state X0 = 1 and thus X1 = 2
for (i in 1:(n - 1))
  x[i + 1] <- r * x[i]
plot(x, type = 'b', xlab = 'time', bty = 'n'
```

Note that we can find any X_t given X_0 by iterating the model as we do in the for loop. X_t is called the solution. Simulation is a bit odd in this case. We can compute the solution analytically. It is $X_t = X_0 r^t$. Thus for $X_{15} = 1 \times 2^{15} = 32768$.

For more complex models, the analytical solution is often not available, and we have to use simulation (the numerical solution).

Note that the exponential model ignores the fact that population growth is limited by resources. At some point food will become scarce. One way of making the model, introduced by Verhulst in 1838, more realistic is to add a growth-limiting term:

$$X_{t+1} = f(X_t) = rX_t \left(1 - \frac{X_t}{K}\right). \tag{2.2}$$

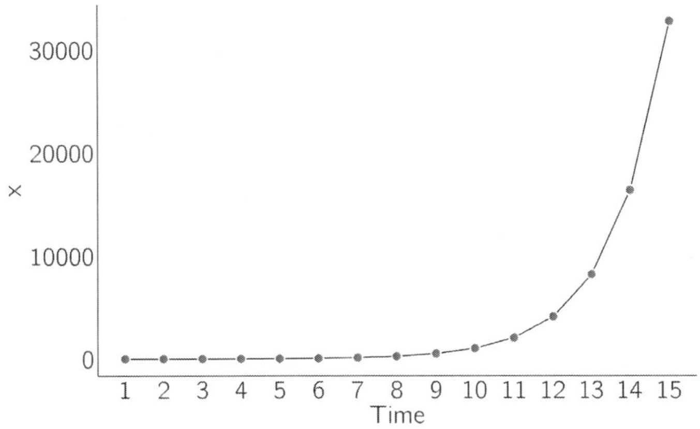

Figure 2.1. Exponential growth.

What is the effect of this addition to the equation? If X is much smaller than the resource K, then the second term, $(1 - X_t/K)$, is close to 1 and we will see exponential growth. But as X approaches K, this term becomes very small, reducing the effect of exponential growth. X does not actually grow up to K, but to a lower value, if it converges at all. We are going to see this in a moment. It also turns out that the actual value of K is not of interest. Changing K does not change the qualitative behavior. Therefore, K is usually set to 1, scaling the population X between 0 and 1. The only remaining parameter is r. Changing r, however, leads to a number of surprising behaviors.[1]

2.3 Stable and Unstable Fixed Points

Let us study a "boring" case first, $r = 2$ (fig. 2.2).

```
n <- 15; r <- 2; x <- rep(0,n)
x[1] <- .01 # initial state
for (i in 1:(n - 1))
  x[i + 1] <- r * x[i] * (1 - x[i])
plot(x, type = 'b', xlab = 'time', bty = 'n')
```

This is the simple case. The population initially develops exponentially but then levels off and reaches a stable state at $X = .5$. We need to

[1]Verhulst proposed this model in the form of a differential equation in continuous time. We will discuss this type of model in chapter 4. In continuous time, nothing particularly spectacular happens, and we only see the kind of behavior displayed in figure 2.2.

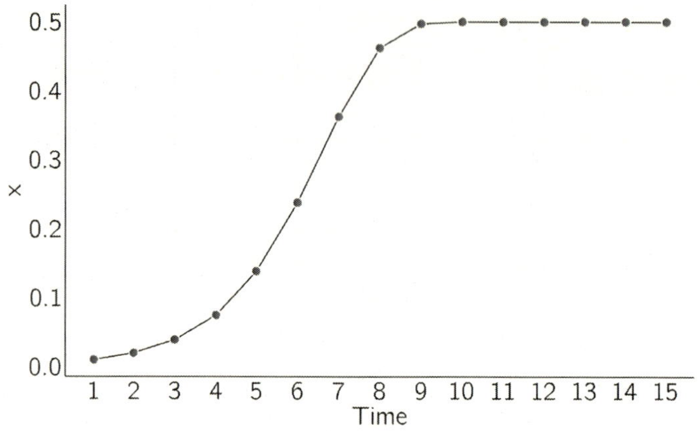

Figure 2.2. The $r = 2$ case. The population converges to a stable state at $X = 0.5$.

understand a bit more about it. What you see here is that we have gone from an unstable initial state to a stable state, a point attractor. The next code shows that this point attractor attracts from a wide range of initial values, but not all.

```
n <- 30; r <- 2; x <- rep(0,n)
for (init in seq(0, .7, by = .01))
  # start from different initial values
{
  x[1] <- init
  for (i in 1:(n - 1))
    x[i + 1] <- r * x[i] * (1 - x[i])
  if (x[i] == 0)
    plot(x,type = 'l',xlab = 'time',bty = 'n',
      ylim = c(0, .8), col = 'red') else
    lines(x)
}
```

If we start exactly at 0, X stays at 0. So, 0 is an equilibrium too, but a special one. It is an unstable fixed point. A small perturbation will cause X to move to .5, the stable fixed point. All initial values in close proximity of 0 will move away from 0 (repellent), but if $X = 0$ exactly, then it remains 0 for all time. So, $X = 0$ is a fixed point but unstable (fig. 2.1).

Fixed points can be stable or unstable.

This concept of equilibrium, stable or unstable, is crucial for later chapters. The essence of the next chapter is to change a control parameter, here r, and study how the pattern of equilibria (the equilibrium

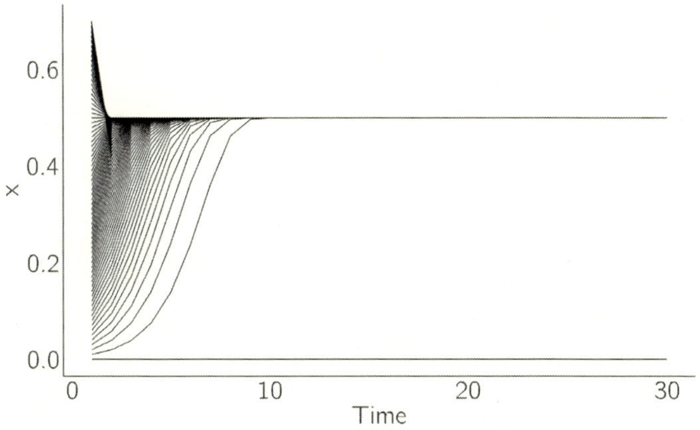

Figure 2.3. Illustration of stable and unstable fixed points. For many initial values, $X = .5$ is an attractor. $X = 0$ is an unstable fixed point. Only if we start exactly at 0 do we stay there.

landscape) changes. You can easily do this yourself by rerunning the simulation with r-values just below and above 1. For $r < 1$, there is only one stable attractor (0).

Simulating this is not really necessary. One has to realize that a fixed point (X^*) is found when $X_{t+1} = X_t = X^*$. See for yourself that:

$$X_{t+1} = X_t = X^*$$
$$X^* = rX^*(1 - X^*)$$
$$X^* = 0 \text{ } or \text{ } 1 = r - rX^*$$
$$X^* = 0 \text{ } or \text{ } X^* = \frac{r-1}{r}.$$

So 0 and $(r - 1)/r$ are fixed points. Indeed, for $r = 2$, we have seen that 0 and .5 are equilibria, one unstable and one stable. To determine whether fixed points are stable, we look at the derivative of the function, $f'(x)$, which, as you can easily check, is $r - 2rX$.

The fixed point is stable if the absolute value of the derivative in the fixed-point value is less than 1.[2] For $r = 2$ the fixed points are 0 and .5. $\left|f'(X^* = 0)\right| = |2 - 0| = 2$, which is greater than 1 and thus $X^* = 0$ is unstable. $\left|f'(X^* = .5)\right| = |2 - 2 \times 2 \times .5| = 0$, which is less than 1 and

[2] It is not too difficult to understand why this is. If you Google search "fixed points of difference equations," you will quickly arrive at stackexchange.com, where several insightful explanations are given.

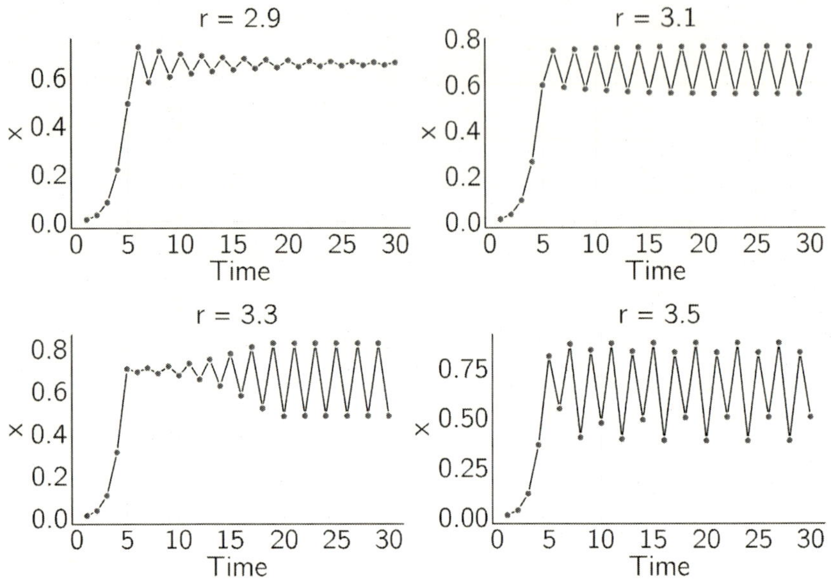

Figure 2.4. Qualitative different behavior of the logistic map for different values of r.

thus $X^* = .5$ is stable. You can check for yourself that $X^* = (r-1)/r$ is stable for $1 < r < 3$, both with the R-code and with the absolute value of the derivative.

2.4 Limit Cycles

So at $r = 3$ the fixed point at $(r-1)/r$ becomes unstable. Let's study some cases. The plots in figure 2.4 are made with the code for figure 2.2.

For $r = 2.9$ we see that the series converges to the fixed point $\frac{1.9}{2.9} = .66$, but in a process of over- and undershooting. Between $r = 3.1$ and $r = 3.3$, a limit cycle of period 2 arises. For $r = 3.5$ this becomes even more remarkable, and we see a limit cycle of period 4. For slightly larger values, we could get cycles with higher periods.

In a limit cycle of period 2, the population oscillates between two values.

It has been claimed that these limit cycles occur in real population dynamics (Hassell, Lawton, and May 1976). Intuitively, it can be understood as a process of over- and undershooting, which dampens out for r a little below 3, but not for $r > 3$.

2.5 Chaos

If we increase r even further, the doubling of the periods changes to even stranger behavior. Figure 2.4 shows what the time series looks for $r = 4$.

There seems to be no regularity. This is what we call deterministic chaos. This time series is unpredictable, even though we know the equation and the system is deterministic. What exactly do we mean by this? Let me illustrate.

```
r <- 4;  n <- 50; x <- rep(0,n)
x[1] <- .001
for (i in 1:(n - 1))
  x[i + 1] <- r * x[i] * (1 - x[i])
plot(x, type = 'l', xlab = 'time', bty = 'n')
# restart with sightly different initial state:
x[1] <- .0010001
for (i in 1:(n - 1))
  x[i + 1] <- r * x[i] * (1 - x[i])
lines(x, col = 'red')
```

We can see that a run with a slightly different initial value will at first follow the same path, but then it will diverge sharply (fig. 2.6). A tiny perturbation (the butterfly flapping its wings) propagates through the system and dramatically changes the long-term course of the system.

This is why long-term weather prediction will never be possible, even if we develop much more precise mathematical models, take more intensive and more accurate measurements, and use more powerful computers.

Note that some uncertainty about the exact value of the initial state is always inevitable. Suppose we have an equation like the logistic map for temperature in the weather system, and this equation perfectly describes that system. To make a prediction, we need to feed the current temperature into the computer. But we cannot measure temperature with infinite precision. And even if we could, we do not have a computer that can handle

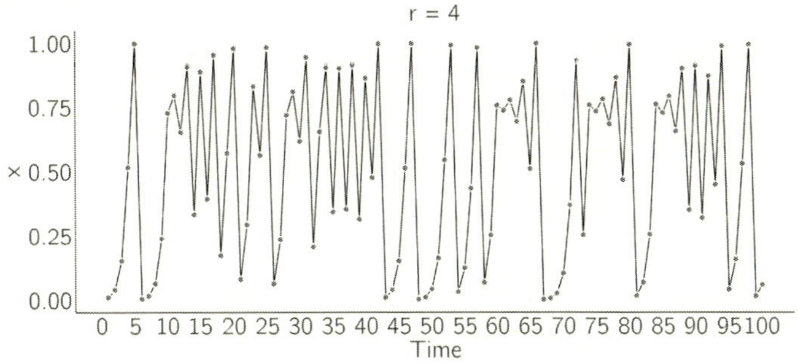

Figure 2.5. Chaos for $r = 4$.

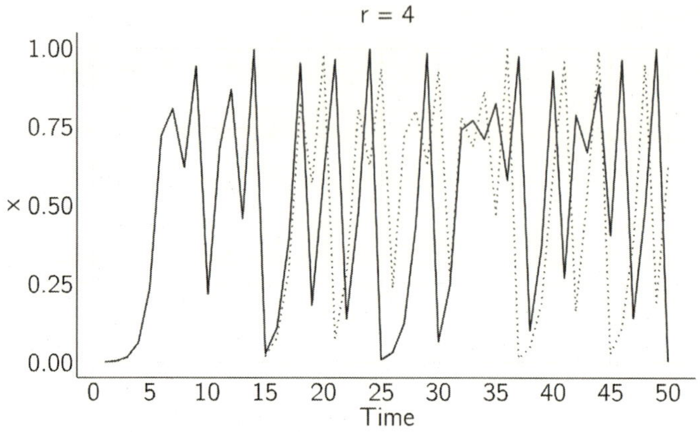

Figure 2.6. The butterfly effect: A small difference in initial state causes divergence in the long run.

numbers with an infinite number of digits. So, we make a small error in setting the initial state, and this will always mess up our long-term forecast. The weather turns out to be a chaotic system. Sensitivity to initial conditions is a necessary and perhaps sufficient condition for deterministic chaos. For a discussion on the definition of chaos, I refer to Banks et al. (1992) and Broer and Takens (2010).

The Lyapunov coefficient quantifies chaos.

The idea of the Lyapunov coefficient is to take two very close initial conditions with a difference of ε. In the next iteration, this difference might be smaller, the same, or bigger. In the last case, the time series diverge, which is typical for chaos. The Lyapunov coefficient is defined as:

$$\lambda_L = \lim_{n\to\infty} \frac{1}{n} \sum_{i}^{n} \ln \left| f^{'}(X_i) \right|. \tag{2.3}$$

where $f^{'}(X_i) = r - 2rX_i$ for the logistic map and $\lambda_L > 0$ indicates chaos. You may verify in a simulation that $\lambda_L > 0$ for $r = 4$, indicating chaos.

2.6 Phase Plot and Bifurcation Diagrams

A phase plot is a graphical representation of the relationship between two or more variables that change over time.

Equation 2.2 is very simple. It is just one equation, a deterministic difference equation specifying how X_{t+1} depends on X_t, but the variety of behavior is astonishing. One way to better understand its behavior is to use phase plots. In one-dimensional systems we plot X_t against X_{t+1} (see fig. 2.7). The code for this figure is:

```
layout(matrix(1:6,2,3))
r <- 3.3; n <- 200; x <- rep(0,n)
x[1] <- .001
for(i in 1:(n-1)) x[i+1] = r*x[i]*(1-x[i])
x <- x[-1:-100]
plot(x, type = 'l', xlab = 'time', bty = 'n',
     main = paste('r = ', r), ylim = 0:1, cex.main = 2)
plot(x[-length(x)], x[-1],
     xlim=0:1, ylim=0:1, xlab='Xt', ylab='Xt+1', bty='n')
r <- 4; x[1] <- .001;
for(i in 1:(n-1)) x[i+1] <- r * x[i] * (1-x[i])
x <- x[-1:-100]
plot(x, type = 'l', xlab = 'time', bty = 'n',
     main = paste('r = ',r), cex.main = 2)
plot(x[-length(x)], x[-1], xlim = 0:1, ylim = 0:1,
     xlab = 'Xt', ylab = 'Xt+1', bty = 'n')
x <- runif(200,0,1)
x <- x[-1:-100]
plot(x, type = 'l', xlab = 'time', bty = 'n',
     main = 'random noise', cex.main = 2)
plot(x[-length(x)],x[-1], xlim = 0:1, ylim = 0:1,
     xlab = 'Xt', ylab = 'Xt+1', bty = 'n')
```

The top figures are time plots, and the lower figures are phase plots. The first column shows a limit cycle of period 2, the second deterministic chaos, and the third noise generated from a uniform distribution. Although the time series of the second and third cases look similar, the phase diagram reveals hidden structure in the chaos time series. Phase plots can help us to distinguish chaos from noise.

The second useful graph is the bifurcation graph. In case of the logistic map, it summarizes the equilibrium behavior for different values of r in one figure. The idea is to plot the equilibria as y-values for a range of r-values on the x-axis. This means that if we take a low r value ($r < 1$), we will only plot 0s, as only $X^* = 0$ is a stable fixed point. Between 1 and 3, we will also see one fixed point equal to $(r-1)/r$. For $r = 3.3$, we expect to see two points as the attractor is a limit cycle with period 2. For higher r we get chaos. How does this all look?

A bifurcation graph is a diagram that shows how the qualitative behavior of a system changes, for example, from stable to chaotic, when one of its parameters changes.

It is actually a good challenge to program this yourself. The trick is to create time series for a range of values of r, delete the first part of this series (we only want the equilibrium behavior), and plot these as y-values. So, if the logistic map has period 2 ($r = 3.3$), we repeatedly plot only two points. For $r = 4$ we get the whole chaos band.

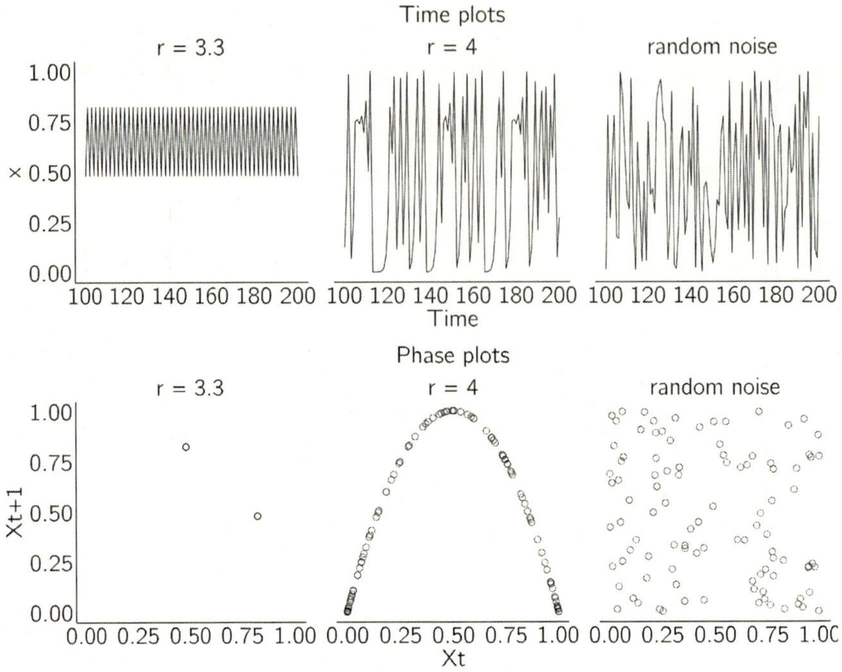

Figure 2.7. Time (top) and phase (bottom) plots for three cases. Chaos and random noise can be distinguished using the phase plots.

A clever way to do this is to use the `sapply` function in R.

```
layout(1)
f <- function(r, x, n, m){
  x <- rep(x,n)
  for(i in 1:(n-1)) x[i+1] <- r * x[i] * (1-x[i])
  x[c((n-m):n)] # only return last m iterations
}
r.range <- seq(0, 2.5, by = 0.01)
r.range <- c(r.range,seq(2.5, 4, by = 0.001))
n <- 200; m <-100
equilibria <- as.vector(
                sapply(r.range, f,  x = 0.1, n = n, m = m-1))
r <- sort(rep(r.range, m))
plot(equilibria ~ r, pch = 19, cex = .01, bty = 'n')
```

Fractals are figures in which certain patterns reappear when we zoom in on the figure, and this happens again and again when we zoom in farther.

This results in figure 2.8. We see indeed fixed stable points for $r < 3$, the period doubling of the limit cycles for $r > 3$, followed by chaos.

Fractals are a recurring phenomenon in many chaotic maps. You can see the fractal nature of the logistic map by zooming in on the interval of

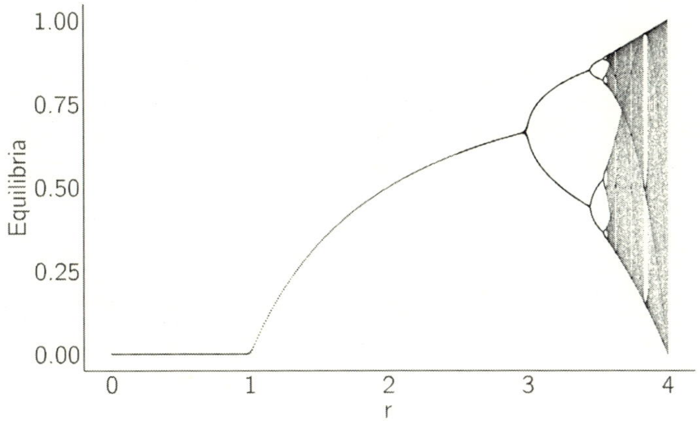

Figure 2.8. The bifurcation diagram of the logistic map.

r between 3.83 and 3.86 (see exercise 2). The three equilibria in the limit cycle split again into period doubling cycles, as we saw in the overall plot between r in 3 and 3.5.

One famous result on this period doubling route to chaos is the Feigenbaum constant. The ratios of distances between consecutive period doubling points (e.g., the distance between first and second divided by the distance between the second and third point) converge to a value of approximately 4.6692. The amazing thing is that this constant is the same for any unimodal map.

2.7 What Did We Learn?

I find these results stunning. I note again that the generating function is deceptively simple, but its behavior is utterly complex and beautiful. Mathematicians have studied every detail of these plots, and most of it is beyond my comprehension. The Wikipedia on the logistic map will introduce you to some more advanced concepts, but for our purposes the present introduction will suffice.

Let's review the concepts we have already learned. The first is the concept of equilibrium. The states of dynamical systems tend to converge to certain values. The simplest of these is the fixed point. Fixed points can be stable or unstable (more on this in the next chapter). If we start a system exactly at its unstable fixed point (and there is no noise in the system), it will stay there. But any small perturbation will cause it to escape and move to the fixed stable point.

The bifurcation diagram summarizes this behavior and also shows how the equilibria change when a control parameter changes. For example, at $r = 1$ we see a bifurcation in the logistic map. Initially 0 was the stable fixed point and $(r - 1)/r$ was unstable. At $r = 1$ this is reversed. At $r = 3$ we see another bifurcation when limit cycles appear.

We have learned that there are all sorts of equilibria. The strangest ones are called strange attractors, which are associated with deterministic chaos. You can see them by making a phase diagram. Phase diagrams for other famous maps are often stunning. The most famous is the Mandelbrot set (look on the internet). There is an R blog about the Mandelbrot set.[3] Simulation helps understanding!

The last thing we learned is that even if our world were deterministic (it is not!), and we knew all the laws of motion (say, the logistic map), and we knew initial states with enormous precision, the world would still be unpredictable.

This statement needs some nuance. I have already mentioned that the weather can be chaotic and unpredictable. But the weather is not always so unpredictable. Sometimes longer forecasts are possible. But forecasts beyond, say, ten days seem out of reach. We also see in the logistic map that when r is close to 4, the forecast suffers from the butterfly effect, but for $r = 2$ the time course is very predictable, even more predictable than in many linear systems. This is because there is only one stable fixed point (.5). The initial state does not matter: we always end up at .5!

The logistic map is either extremely predictable or extremely unpredictable depending on the value of r.

2.8 Other Maps and Fractals

There are many accessible sources on chaos theory. As always, Wikipedia is a great resource. It helps me a lot by actually doing things, that is, doing computer simulations. One example is the Henon map, which consists of two coupled difference equations:

[3] https://www.r-bloggers.com/2017/06/the-mandelbrot-set-in-r-2/. Be sure to check the Shiny app.

$$X_{t+1} = 1 - aX_t^2 + Y_t,$$
$$Y_{t+1} = bX_t. \tag{2.4}$$

Using the code example from the logistic map, you should be able to generate time series and a phase diagram for this model. Try to reproduce the first image of the Wikipedia page on the Henon map. The amazing three-dimensional bifurcation diagram may be more challenging.

Fractals are another topic for further study. Another look at Wikipedia is recommended. Making your own fractals in R is made easy by the R blog by Martin Stefan (2020).

2.9 Detecting Chaos in Psychophysiological Data

Chaos theory and the logistic map were popularized about fifty years ago, and since then researchers have been looking for chaos in all kinds of time series (Ayers 1997; Robertson and Combs 2014; G. K. Schiepek et al. 2017). One idea behind this work is the hypothesis that chaos might be healthy (Pool 1989) or helpful. It would be helpful in learning algorithms, such as neural networks, to prevent getting stuck in local minima (Bertschinger and Natschläger 2004). My very first publication was about chaos in neural networks (van der Maas, Verschure, and Molenaar 1990).

There are many techniques for chaos detection in times series. However, these empirical signals are inevitably contaminated with noise (Rosso et al. 2007). One example is the computation of Lyapunov exponents, quantifying how small differences in initial conditions evolve over time. A positive Lyapunov exponent indicates chaos, signifying exponential divergence of trajectories, which is a hallmark of chaotic systems. This method involves reconstructing the phase space from time-series data and calculating the average exponential rate of separation of trajectories. With the Lyapunov function in the package DChaos, you can compute the Lyapunov coefficient for times series generated with the logistic map. You may verify that for $r = 4$, you get the Lyapunov coefficient as computed with the derivative earlier. There are several packages available in R, including new methods based on machine-learning techniques (Sandubete and Escot 2021; Toker, Sommer, and D'Esposito 2020).

Chaos detection is an active area of research, with new methods being proposed on a regular basis (Zanin 2022).

These methods generally require long time series. Many publications appeared on the detection of chaos in psychophysiological data. Examples are electroencephalogram (EEG) (Pritchard and Duke 1992), heartbeat (Freitas et al. 2009), electromyogram (EMG) (Lei, Wang, and Feng 2001), and eye movements (Harezlak and Kasprowski 2018). Reviews of these lines of research are provided by Stam (2005), Kargarnovin et al. (2023), and Garc and Pe (2015).

2.10 Exercises

1. For $r = 3.5$, the logistic map iterates between four points. For which value(s) does it iterate between 8 points? (*)

2. Section 2.6 shows the code to make a bifurcation plot. First, run this code and look at the bifurcation plot. In this plot, you can also zoom in by changing the interval between the rs on the x-axis. Adjust the code by changing the `r.range` to `seq(3.4, 4, by = 0.0001)` , also change `cex = 0.01` to a lower value. Zoom in on the interval of r between 3.83 and 3.86. In this interval the chaos suddenly disappears and limit cycles with period 3 appear. Check this with a time-series plot for a particular value of r. (*)

3. Reproduce the first image from the Henon map Wikipedia page. Provide your R code and figure. (*)

4. Make the bifurcation diagram of the Ricker model (see Wikipedia). Provide your R code and figure. Why is this model considered a more realistic representation of population growth than the logistic map? (*)

5. Also reproduce the three-dimensional bifurcation diagram of the Henon map. (**)

6. Have a look at the definition of the Lyapunov coefficient in section 2.5. Calculate this Lyapunov coefficient for the logistic map where $r = 4$ using the Dchaos package in R. This coefficient can also be calculated manually using the derivative (eq. 2.3). Do this and check that the coefficients are approximately equal. (**)

7. Use the Rmusic library installed with `devtools::install_github("keithmcnulty/Rmusic", build_vignettes = TRUE)` to create a chaos sound machine. Make one for white noise too. Can you hear the difference? (**)

8. Find a paper on chaos detection in psychology or psychophysiology and summarize it in 300–400 words. (*)

Ω

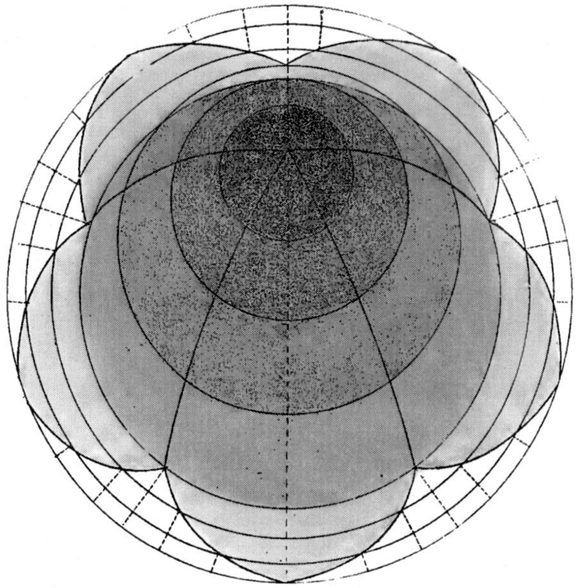

$\genfrac{}{}{0pt}{}{D}{G}\phi^{2}\chi^{5}_{3}$

A^{1}

A

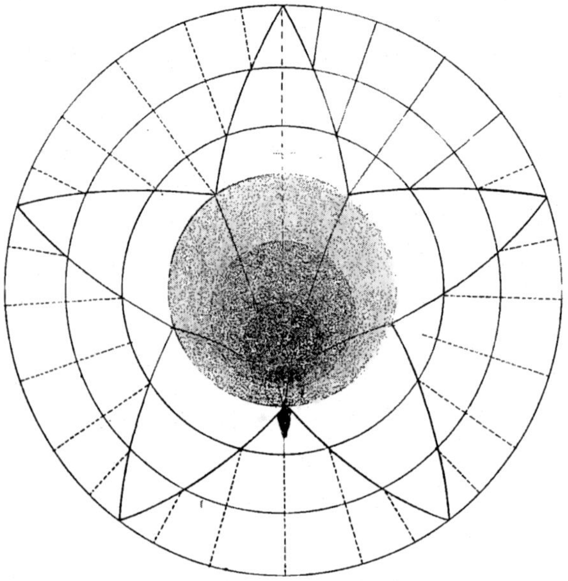

3

TRANSITIONS IN COMPLEX SYSTEMS

3.1 Introduction

My dissertation research was on Jean Piaget's stage theory of cognitive development. These stages were separated by transitions. One such transition should occur between the pre-operational and concrete-operational stages. In the concrete-operational stage, children learn logical, concrete physical rules about objects, such as weight, height, and volume. The most famous test to distinguish between the two stages is the conservation task.

RESOURCES
The electronic version of this book, including code blocks and links, can be viewed and accessed at: *santafeinstitute.github.io/ComplexPsych/*

There are many conservation tasks, but the setup is always the same. For example, you show a child two equal balls of clay, ask for confirmation that they weigh the same, roll one into a sausage shape, and then ask again for confirmation of equal weight. A nonconserving child will now claim that the longer sausage weighs more. One can also do this with two rows of coins (spreading one row out) or two glasses of water (pouring the water from one glass into a smaller longer glass). It is actually a fascinating task to do with children between five and eight years old.

From the 1960s to the 1980s, this was a topic of major interest in developmental psychology. A key question was whether there really was a stage transition, and there was a lot of confusion about what a transition actually was. It was my task to clarify this and to prove Piaget's hypothesis. I think I succeeded in clarifying the question, but whether I succeeded in proving the stage theory is debatable.[1]

My PhD advisor Peter Molenaar had the idea to use catastrophe theory to define the concept of a transition in a precise way, to use the so-called catastrophe flags to test the hypothesis of a transition, and also to fit a cusp model to the conservation data. It took me, with the help of many people, more than twenty years to do all these steps (van der Maas and Molenaar

[1] Learning a particular conservation task does seem to be rather sudden, but there could easily be two years between learning conservation of number and conservation of volume (Kreitler and Kreitler 1989). This is inconsistent with the stage theory.

1992; Jansen and van der Maas 2001; Dolan and van der Maas 1998; Grasman, van der Maas, and Wagenmakers 2009).

What is catastrophe theory, what are these flags, and what is the cusp? These are the first questions I will answer in this chapter. But this chapter is also about statistics. You will learn how to fit a cusp model to data. I will present a methodology for studying transitions in areas where we do not have a mathematical description of the underlying system. I will present examples from very different subfields of psychology. Finally, I will discuss the criticisms that were made in response to the hype around catastrophe theory about fifty years ago. This is a long chapter, but I have tried to include only what is necessary in order to make intelligent use of this approach. This requires some basic understanding of the mathematics of catastrophe theory, a good overview of the possibilities for testing cusp models, and knowledge of the controversies from the early days of the popularization of this theory.

3.2 Examples of Transitions

In chapter 1, section 1.5, I stated that complex systems tend to be characterized by a limited number of equilibria. As we saw in the chapter 2, these equilibria can take many different forms, but in this chapter, we consider only stable and unstable fixed points. We are particularly interested in the case where the configuration of stable and unstable points changes due to a smooth change in some external variable, a control variable. In such a case a discontinuous change or a (first-order) phase transition can occur. A transition or tipping point is an intriguing property of complex systems.

I call it an intriguing property because in the linear systems we are used to, smooth changes in control variables lead to similar (proportional) changes in behavior variables. We may see a big change in some behavior, but this requires a big change in the controls. An example would be the speed of your bike and the force you apply. But in the case of fear or panic, this process is often nonlinear.

If a smooth change in an independent or control variable, such as the smell of smoke, leads to a sudden jump in fear (e.g., panic), we are likely to be dealing with a phase transition.

A key physical example is the change in state of water. Between, say, 10 and 80 degrees Celsius, a smooth change in temperature results in only a slight change in the liquid state of water. But if we change the temperature very slowly, close to the thresholds of 0 or 100 degrees Celsius, we see sudden phase transitions.

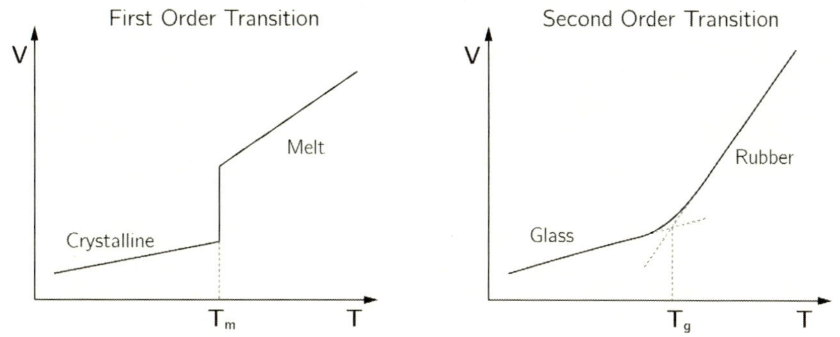

Figure 3.1. Examples of a first-order (discontinuous) and second-order (continuous) phase transition. In these particular cases, V represents volume and T temperature.

We saw something similar for the logistic map when r crossed the boundary at $r = 1$. However, this did not lead to a sudden change in X^*. This is often called a second-order phase transition, meaning that the configuration of stable and unstable points changes, but there is no discontinuous change in behavior (fig. 3.1).[2]

Discontinuous phase transitions such as melting and freezing occur in many systems. Famous examples from the natural sciences include collapsing bridges, capsizing ships, cell division, and climate transitions such as the onset of ice ages. Examples from the social sciences include conflict, war, and revolution. Some examples from psychology are falling asleep, outbursts of aggression, radicalization, falling in love, sudden insights, relapses into depression or addiction, panic, and multistable perception. The perception of the Necker cube is a famous example (fig. 3.2). Building and testing models of these psychological transitions is challenging but rewarding. These transitions involve large changes in behavior, in contrast to the smaller, often marginally significant effects typically observed in psychological intervention studies.

3.3 Bifurcation and Catastrophe Theory

Bifurcations occur when equilibria disappear, appear, or split. Simply put, bifurcation theory studies how small changes in parameters or conditions can lead to large changes in outcomes in mathematical systems.

Bifurcation theory is a branch of mathematics that studies changes in the qualitative or topological structure of a given family of dynamical systems as parameters are smoothly varied.

[2] A related and very similar distinction is that between a subcritical bifurcation and a supercritical bifurcation.

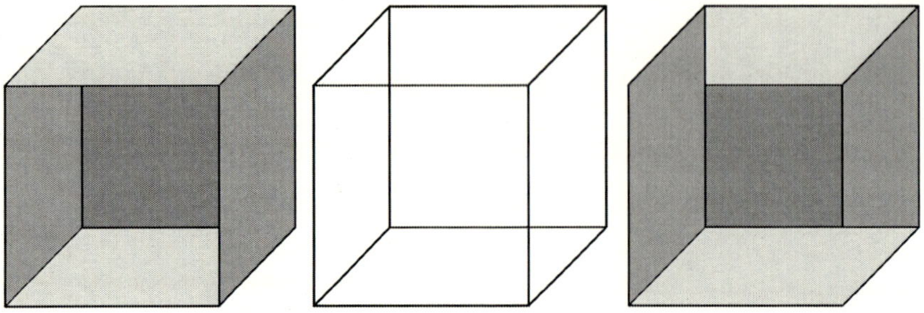

Figure 3.2. Transitions in the perception of the Necker cube. The perception of the middle cube is bistable, and sudden transitions occur between the left ("front") and right ("back") percepts. Multistable perception is a much-studied psychological phenomenon that is still not fully understood.

Catastrophe theory can be viewed as a branch of bifurcation theory, describing a subclass of bifurcations. It was developed by René Thom (1977) and popularized by Christopher Zeeman (1976). The reason I chose to focus on catastrophe theory in this chapter is fourfold: First, it provides one of the few systematic treatments of bifurcations. A systematic treatment is more effective than simply listing all types of bifurcations. Second, once you have a grasp of the basics of catastrophe theory, it becomes easier to learn about other bifurcations not encompassed by this theory. Third, it is the most widely used approach in psychology and the social sciences. Finally, the field has developed an empirical program and statistical procedures for the practical application of catastrophe theory.

In gradient systems some quantity, such as energy, is minimized or maximized.

Catastrophe theory is concerned with gradient systems. These are dynamic systems that can be described by a potential function. Potential functions can be thought of as landscapes with minima and maxima in which we throw a ball and see where it ends up. The simplest case, discussed in the next section, is the quadratic minimum. We can also study what happens to the ball if the landscape changes shape smoothly and a minimum disappears. Then sudden jumps can occur.

A potential function in mathematics describes the potential energy landscape of a system, where the system's dynamics are determined by the gradients of this function.

Minima and maxima are called critical points, points where the first derivative of the potential function is 0. Catastrophe theory analyzes so-called degenerate critical points of the potential function. Phase transitions can occur at these bifurcation points. Thom proved that there are only seven fundamental types of catastrophes (given a limited set of control parameters). I will start with a mathematical introduction and, after

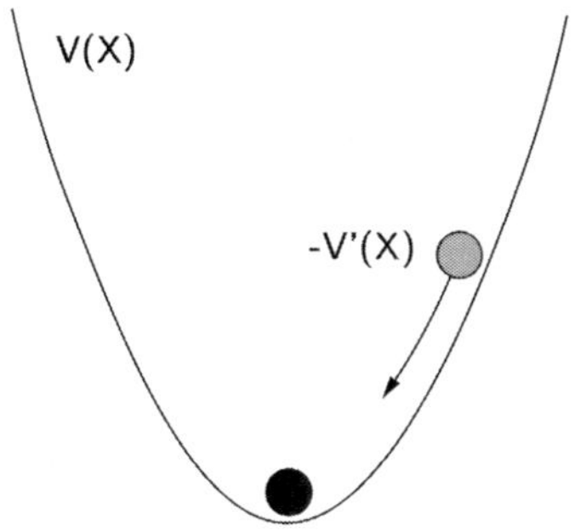

Figure 3.3. The quadratic potential function. A ball rolls to the minimum value of $V(X)$. Its change is defined by the negative of the derivative of the potential function $-V'(X)$.

explaining the main concepts, give some psychological examples. An in-depth discussion of the role of potential functions in catastrophe theory can be found in the introduction of chapter 1 of Gilmore (1993).

Degenerate critical points are points where not only the first derivative but also the second derivative of the potential function is zero.

3.3.1 THE QUADRATIC CASE

Thom's theorems are known to be highly complicated, but the basic concepts are not that difficult to grasp. The simplest potential function is

$$V(X) = X^2. \tag{3.1}$$

Imagine a ball in a landscape. The ball will roll to the minimum of the potential function (figure 3.3). We learned in school that this is the point where the first derivative is 0 and the second derivative is positive. The first and second derivatives are $V'(X) = 2X$ and $V''(X) = 2$, respectively. At $X = 0$ we find the minimum.

The potential function describes a dynamical system defined by

$$\frac{dX}{dt} = -V'(X). \tag{3.2}$$

This makes sense. When the ball is in $(1, 1)$, $-V'(X) = -2$ and the ball will move toward $X = 0$. But if $X = 0$, $-V'(X) = 0$, and the ball will not move anymore. In the case of the quadratic potential function, there is only one fixed point. By adding parameters and lower order terms to V, that is, $aX + X^2$, we can move its location, but the qualitative form (one stable fixed point) will not change. Also note that

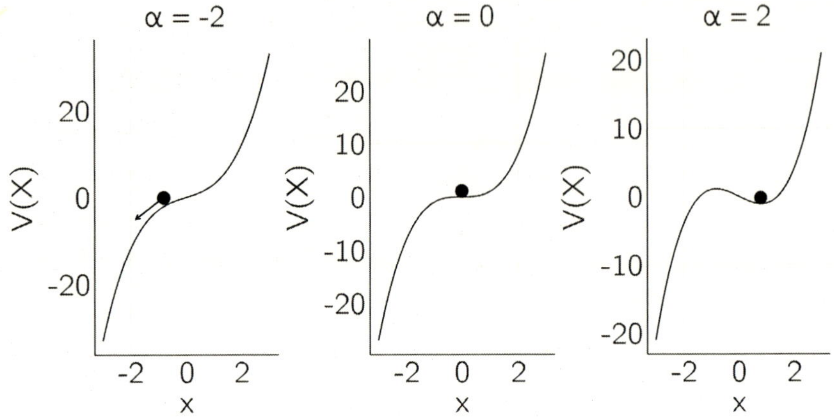

Figure 3.4. A bifurcation at $a = 0$: the equilibria change qualitatively. For $a < 0$ there is no equilibrium; for $a > 0$ we have a minimum and a maximum.

the second derivative is positive, which tells us that we are dealing with a minimum and not a maximum (the so-called second derivative test).

Many dynamical systems behave according to this potential function.[3] Nothing spectacular happens: no bifurcations and no jumps. This is different when we consider potential functions with higher-order terms.

3.3.2 THE FOLD CATASTROPHE

The fold catastrophe is defined by the potential function

$$V(X) = -aX + X^3. \tag{3.3}$$

This function has a degenerate critical (bifurcation) point at $X = 0, a = 0$, because at this point $V^{'}(X) = -a + 3X^2 = 0$ and $V^{''}(X) = 6X = 0$, so both the first and second derivative are 0. What makes this point so special? This is illustrated in figure 3.4.

```
layout(t(1:3))
V <- function(X,a) -a * X + X^3
curve(V(x, a = -2), -3, 3, bty = 'n')
curve(V(x, a = 0), -3, 3, bty = 'n')
curve(V(x, a = 2), -3, 3, bty = 'n')
```

[3] As we will see in section 3.5, the statistical equivalent of the quadratic potential function is the normal distribution, the most popular distribution in our statistical work.

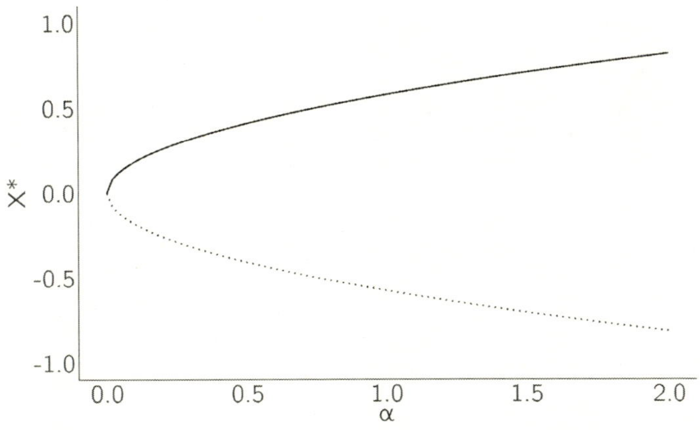

Figure 3.5. The bifurcation diagram of the fold catastrophe. Similar to what's shown in figure 3.4, when $a = 0$, there is a dramatic change in the equilibrium landscape. Suddenly, both a stable and an unstable equilibrium emerge, seemingly from nowhere.

In the left plot, $a < 0$ and there is no fixed point; the ball rolls away to minus infinity. This can be checked by setting the first derivative to zero, which gives $X = \pm\sqrt{a/3}$. For negative a there is no solution. A positive value of a gives two solutions, as shown on the right for $a = 2$. The positive solution $X = \sqrt{2/3}$ is a stable fixed point because the second derivative in this point is positive. The negative solution $X = -\sqrt{2/3}$ is an unstable fixed point because the second derivative in this point is negative.

The middle figure depicts the case just in between these two cases. Here the equilibrium is an inflection point, a degenerate critical point. The bifurcation occurs at this point as we go from a landscape with no fixed points to one with two, one stable and one unstable.

Another way to visualize this is by making a bifurcation diagram as we did for the logistic map in chapter 2. On the x-axis we put a, from 0 to 2. On the y-axis we plot X^*, the fixed points of equation 3.3. We use lines for stable fixed points and dashed lines for unstable points. The diagram is shown in figure 3.5.

Note that the fold is not accompanied by sudden jumps in behavior. It is an example of a second-order phase transition.

This bifurcation diagram may not look as spectacular as the logistic map, but its importance cannot be overstated. The fold is everywhere! In a fascinating book, *The Seduction of Curves*, Allen McRobie (2017) shows that whenever we see an edge, we see a fold. Figure 3.6 is from the

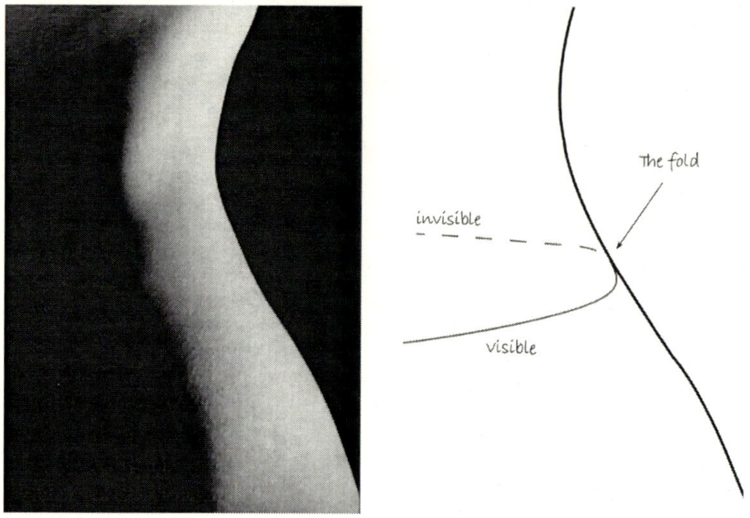

Figure 3.6. The fold in drawings. The fold line separates the parts that can be seen from the parts that are hidden. *(Adapted from from McRobie (2017) with permission.)*

book, where he demonstrates how different catastrophes appear in art. I also recommend his YouTube lecture.[4]

The fold catastrophe is also known as a saddle node, tangential, or blue-sky bifurcation.

The fold catastrophe has been studied in fields from evolution theory (Dodson and Hallam 1977) to buoyancy in diving (Güémez, Fiolhais, and Fiolhais 2002). In addition, higher-order catastrophes are composed of folds.

3.3.3 THE CUSP CATASTROPHE

Sudden jumps between stable states are associated with first-order phase transitions.

The cusp, the best-known catastrophe, is the simplest catastrophe showing sudden jumps in behavior. The potential function of the cusp is

$$V(X) = -aX - \frac{1}{2}bX^2 + \frac{1}{4}X^4. \tag{3.4}$$

Control variables are the parameters whose gradual changes induce qualitative change in the behavior of the system.

The half and quarter are added to make later derivations a little easier. The highest power is now 4. The first two terms contain the control variables a and b, known as the normal and splitting variables. You might ask why there is no third-order term. The nontechnical answer is that such a term would not change the qualitative behavior of the bifurcation. Catastrophe theory studies bifurcations that are structurally stable, meaning that

[4] https://www.youtube.com/watch?v=6ZQKzcw9Ulk

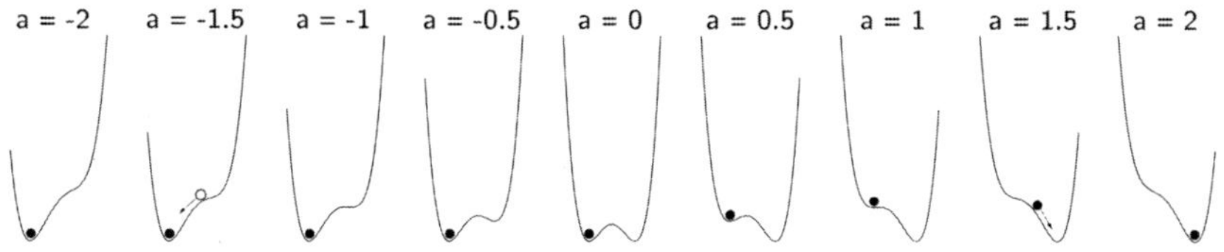

Figure 3.7. The change in the potential function of the cusp by varying a. Note that the jump to the other state does not happen at $a = 0$ but is delayed and depends on the direction of the change in a. This delay is called hysteresis.

perturbing the equations (and not just the parameters) does not fundamentally change the behavior (see sec. 3.3 and Stewart (1982) for further explication).

I advise you to do some minimal research on this equation yourself, using an online graphic calculator tool like Desmos or GeoGebra (paste f(X)=-a X-(1/2) b X^2+(1/4) X^4). For example, set $a = 1$ and $b = 3$ and look at the graph of the potential function. Think in terms of the ball moving to a stable fixed point. What you should see is that there are three fixed points, of which the middle one is unstable. This bistability is important. Again, there is a relationship to unpredictability. Although you know the potential function and the values of a and b, you are still not sure where the ball is. It could be in either of the minima.

Other typical behavior occurs when we slowly vary a (up and down from -2 to 2), for a positive b value ($b = 2$). This is shown in figure 3.7. At about $a = 1.5$ we see the sudden jump. The left fixed point loses its stability and the ball rolls to the other minimum.

Now consider what will happen if we decrease a from 2 to -2. In this case, the ball will stay in the right minimum until $a = -1.5$. Where the jump takes place depends on the direction of the change in a, the normal variable. The delay in jumps is called hysteresis. Hysteresis is of great importance in understanding change or lack of change in complex systems. The state in which the ball is the less deep minimum (for $a = 0.5$ in figure 3.7) is often called a metastable or locally stable state.

Hysteresis means lagging behind, or resistance to change.

Metastable states appear to be stable for some time but are not in their globally stable state.

In his classic paper on the psychophysical law, Stevens (1957) reports hysteresis in perceptual judgments when properties such as brightness and loudness are systematically varied from low to high and vice versa. In this paper he says: “I’m trying to describe it, not explain it. I am not sure I know

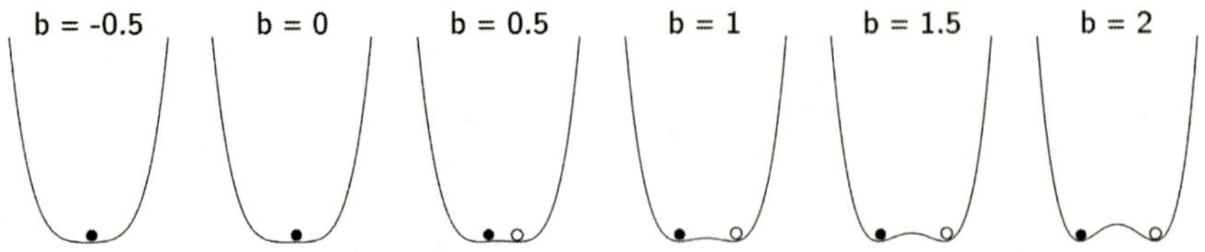

Figure 3.8. The change in the potential function of the cusp by varying b. One minimum splits up in two.

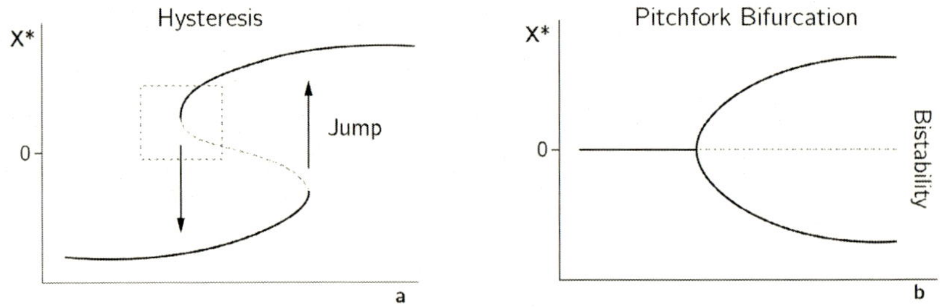

Figure 3.9. Bifurcation plots for the a and b parameters of the cusp. Moving along the a-axis, assuming b is positive, gives hysteresis. Moving along the b axis, assuming $a = 0$, gives the pitchfork bifurcation or divergence. The dotted lines represent unstable maxima. The area in the dotted box in the first plot is a fold.

how to explain it." To me the cusp at least partially explains why hysteresis occurs.

Gilmore (1993) made an important point about noise in the system. If there is a lot of noise, the jumps occur earlier and we see less or no hysteresis effect. This is called the Maxwell convention as opposed to the "delay" convention. Demonstrating hysteresis therefore requires precise experimental control.

Another very interesting pattern occurs when $a = 0$ and b is increased (figure 3.8).

A pitchfork bifurcation occurs when a single equilibrium splits into three (two stable and one unstable) as a parameter changes, resembling a pitchfork's shape.

For low b there is one stable fixed point that becomes unstable. It splits up into two new stable attractors. As we did for the fold, we can make bifurcation diagrams showing the equilibria of X as a function of a and b. Along the a-axis we see hysteresis and along the b-axis we see divergence or what is often called a pitchfork bifurcation (fig. 3.9).

Depicting the combined effects of a and b requires a three-dimensional plot, which combines the hysteresis and pitchfork diagrams (fig. 3.10).

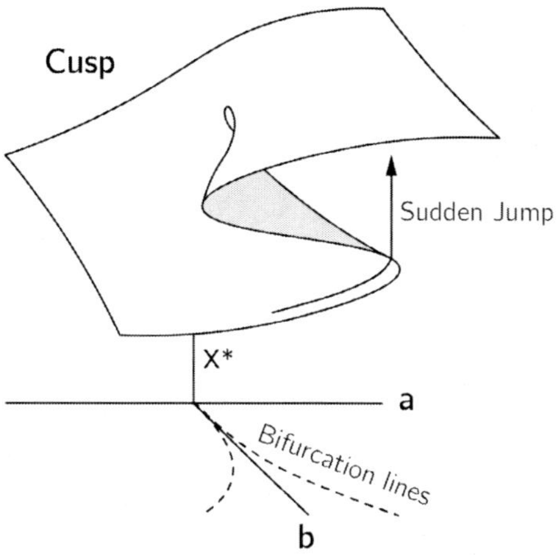

Figure 3.10. The cusp catastrophe combines hysteresis along the normal axis (a) and the pitchfork along the splitting axis (b). At the back of the cusp, changes in a only lead to smooth changes in the equilibrium behavior X^*. At the front, sudden jumps occur when we cross the bifurcation lines. These jumps are typical of first order phase transitions. The area between the bifurcation lines is called the bifurcation set. In this area there are two stable and one unstable equilibrium (shaded gray).

The cusp diagram can be expressed mathematically by setting the first derivative to 0:

$$V^{'}(X) = -a - bX + X^3 = 0. \tag{3.5}$$

This type of equation is called a cubic equation.[5] The degenerate critical points of the cusp can be found by setting the first and second derivative to 0. This is just within reach of your high school mathematics training, and I leave this as an exercise. The result is:

$$27a^2 = 4b^3. \tag{3.6}$$

This equation defines the bifurcation lines where the first and second derivatives are both 0 and sudden jumps occur (see fig. 3.10). The region between the bifurcation lines is the bifurcation set. In this region, the cusp has three fixed points, the middle of which is unstable. These unstable states in

[5]The cubic equation cannot be solved easily. This is due to the fact that the cusp is not a function of the form $y = f(x)$. Functions assign to each element of x exactly one element of y. But in bistable systems we assign two values of y to one value of x.

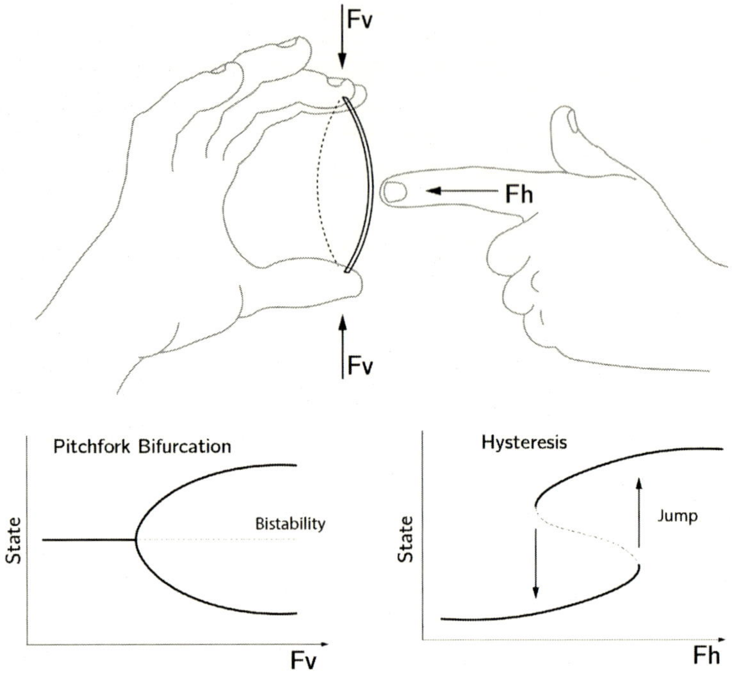

Figure 3.11. A simple business card can be used to illustrate all the properties of the cusp (see main text).

the middle are called the inaccessible area, shaded gray in the cusp diagram. The bifurcation lines meet at $(0, 0, 0)$. At this point, the third derivative is also 0. This is the cusp point.

3.3.4 EXAMPLES OF CUSP MODELS

To illustrate the cusp, I always use the business card (fig. 3.11). I recommend that you test this example (not with your credit card). You can play with two forces. Fv is the vertical force and the splitting control variable (b) in the cusp. Fh is the horizontal force and the normal variable (a) in the cusp. Note that you will only get smooth changes when $Fv = 0$, but sudden jumps and hysteresis when you employ vertical force. One very important phenomenon is that the card has no "memory" when $Fv = 0$. You can push the card to a position, but as soon as you release this force (Fh back to 0), the card moves back to the center position. This is not the case with vertical pressure. If we force the card to the left or right position, it will stay there, even if we remove the horizontal force. The card has a memory.

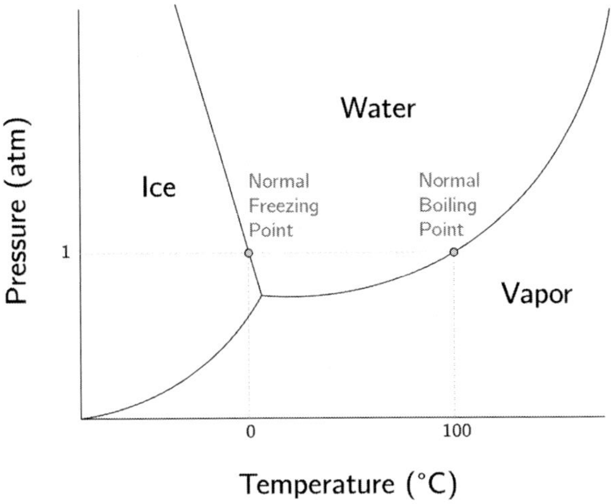

Figure 3.12. The phase diagram of water, which summarizes the equilibria and transitions in the state of water as a function of temperature and pressure.

This seems simple, but the mathematical analysis of such elastic bending structures is a huge topic in itself (Poston and Stewart 2014). The freezing of water is also a cusp. As an approximation, we could say that the density of water is the behavioral variable, temperature is the normal variable, and pressure acts as a splitting variable (see ch. 14 of Poston and Stewart (2014), for a more nuanced analysis). It is very instructive to study the full phase diagram of water (fig. 3.12). It can be viewed as a map of the equilibria. This type of mapping would be extremely useful in psychology and the social sciences.

A psychological example of a cusp concerns sudden jumps in attitudes (Latané and Nowak 1994; van der Maas, Kolstein, and van der Pligt 2003). Attitudes will be discussed in much more detail in later chapters. In general, we have relatively stable attitudes toward many things in life (politics, snakes, hamburgers, and sports), but sometimes they change, and in rare cases they change radically. For example, you may suddenly become a conspiracy theorist, an atheist, or a vegetarian. One example is the attitude toward abortion (fig. 3.13).

Cusp modeling begins by defining the states of the behavioral variable. In this case, the two states of the bistable cusp are the two opposing posi-

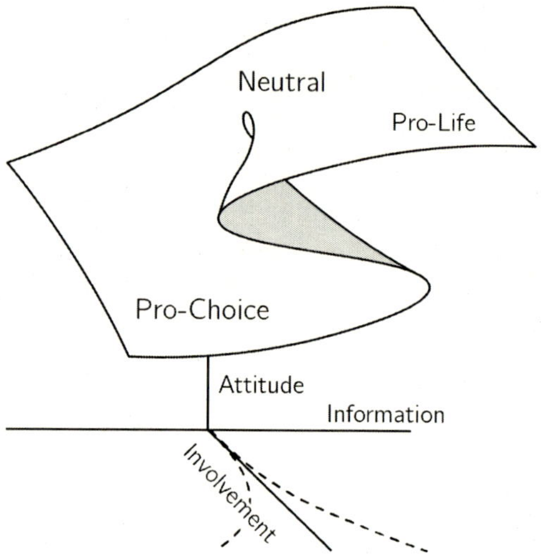

Figure 3.13. The cusp model of attitudes (here toward abortion). Because of hysteresis, it is very difficult to persuade highly involved people with new information, but if they change it will be a sudden jump.

tions, pro-life[6] and pro-choice. The other state associated with $a = 0, b = 0$ is the neutral state, an "I don't know" or "I don't care" position.

The normal (a) and splitting (b) variable are interpreted as information and involvement. Information is a collection of factors that influence whether people tend to be in the pro-life or pro-choice position. Political and religious orientation as well as personal experiences add to this overall factor. One way to construct this information variable is through a factor analysis or principal component analysis.

The splitting factor, involvement, also combines a number of effects (importance, attention). The main idea is that there are two types of independent variables. Some will work (mainly) along the normal axis, and some will (mainly) impact the splitting axis.

The implications of this model are that, for low involvement, change is continuous (fig. 3.13). Presenting people with some new information supporting the pro-life or pro-choice position will have a moderate effect. One problem, as demonstrated with the business card, is that the uninvolved person has "no memory." As soon as you stop influencing this person, they drift to the neutral "I don't care" position. We have another problem when

[6]Perhaps better to call this view anti-choice.

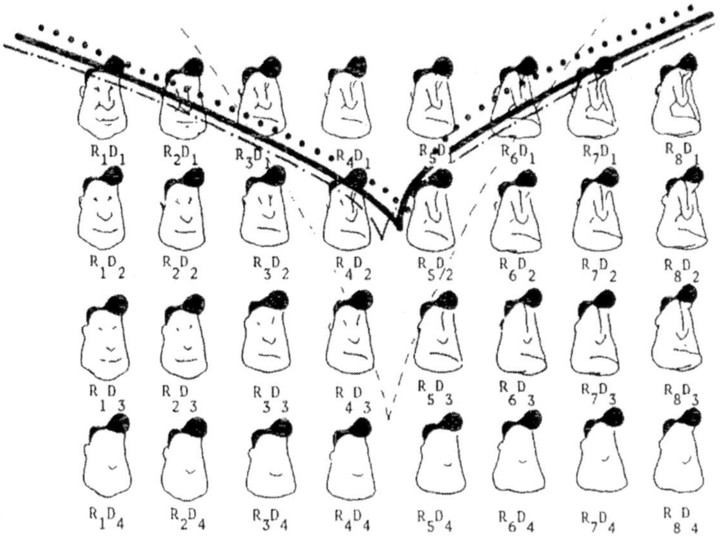

Figure 3.14. Multistable perceptual stimuli positioned in the bifurcation set. The fitted bifurcation lines were calculated using Cobb's method, which is explained in section 3.5. *(Adapted from Stewart and Peregoy 1983 with permission.)*

people are highly involved: they experience hysteresis. When this hysteresis effect is large, persuasion just does not work. If you have been involved in political discussions, you have probably experienced that yourself.

Because of the hysteresis effect, it is very difficult to persuade people with new information.

But if the underlying change in information is large enough, attitudes can show a sudden jump. If they are central attitudes, they can be major life events. There is a lot of anecdotal evidence for such transitions (Ebaugh 1988), but it is very hard to capture such an effect in actual time series of attitude measures. Another effect that is consistent with the cusp model is ambivalence. Ambivalence is associated with high involvement. Highly involved people with balanced information ($a = 0$), may oscillate between extreme positions (see fig. 4.2). Finally, the pitchfork bifurcation can explain issue or political polarization.

In the cusp model of attitudes, ambivalence is not the same thing as being neutral.

Another psychological example of the cusp-like behavior is multistable perception. Stewart and Peregoy (1983) proposed a model in which the perception of male face or female figure is used as a behavioral variable, the splitting variable is the amount of detail, and the normal variable is a change in detail related to the male/female distinction. The results are shown in figure 3.14.

When involvement increases in a group of neutral people, for example, due to discussion, they may split into two extreme positions (polarization).

3.3.5 HIGHER-ORDER CATASTROPHES

Note that the cusp is made up of folds. This is best seen in the hysteresis diagram in figure 3.9 (see the dotted rectangle). Higher order catastrophes yield elements of cusps and folds. The swallowtail catastrophe with potential function $V(X) = -aX - \frac{1}{2}bX^2 - \frac{1}{3}cX^3 + \frac{1}{5}X^5$ consists of three surfaces of fold bifurcations meeting in two lines of cusp bifurcations, which in turn meet in a single swallowtail bifurcation point. We need a four-dimensional space to visualize this, which is difficult. The Wikipedia page on catastrophe theory has some rotating graphs that may help. The butterfly catastrophe has X^6 as the highest term (and four control variables).

The butterfly catastrophe is of interest when we observe trimodal behavior.

I will discuss this catastrophe in section 6.3 in relation to modeling attitudes.[7] Other catastrophes have two behavioral variables, not one. However, the vast majority of applications of catastrophe theory focus on the cusp, which will also be the focus of the remainder of this chapter. There are many good (but not easy) books that present the full scope of catastrophe theory (Gilmore 1993; Poston and Stewart 2014).

3.3.6 OTHER BIFURCATIONS

In contrast to bifurcation theory, catastrophe theory is limited to structurally stable, local bifurcations.

Bifurcation theory also deals with nonstructurally stable bifurcations and so-called global bifurcations.

Examples of nonstructuraly stable local bifurcations are the transcritical bifurcation ($\frac{dX}{dt} = aX - X^2$) and pitchfork bifurcation ($\frac{dX}{dt} = bX - X^3$). The pitchfork is part of the cusp and is not structurally stable because it can be perturbed by an additional term a, which, if unequal to 0, will distort the pitchfork (see fig. 3.15).

Another one we have already seen is the period doubling bifurcation. This happened in the logistic map when the fixed point changed in a limit cycle of period 2. Finally, global bifurcations cannot be localized to a small neighborhood in the phase space, such as when a limit cycle diverges (Guckenheimer and Holmes 1983). However, I don't know of any applications of global bifurcations in psychology or the social sciences.

---

[7] I note that the butterfly catastrophe and the butterfly effect in chaos theory are completely unrelated concepts.

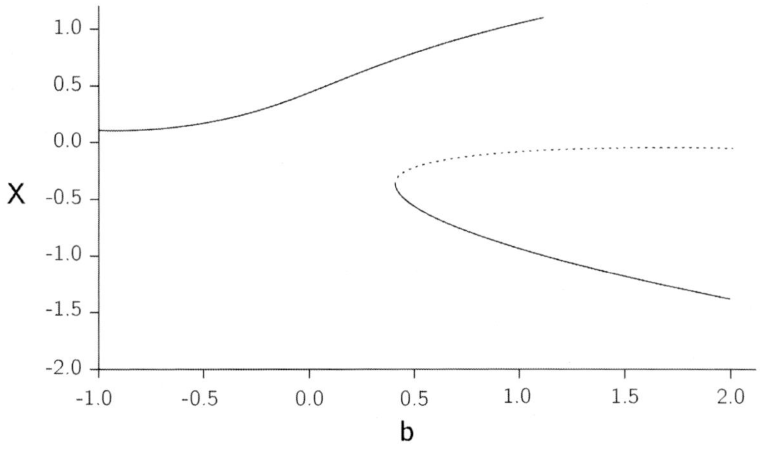

Figure 3.15. Perturbed pitchfork bifurcation ($a = .1$). For $a = 0$ we would get the pitchfork bifurcation as shown in figure 3.9. Thus, a perturbation in a model parameter leads to qualitative change in this bifurcation, and this is why it is not considered structurally stable.

3.4 Building Catastrophe Models

3.4.1 MECHANISTIC MODELS

The model of the attitude toward abortion is called a phenomenological model, as opposed to a mechanistic model.

In phenomenological models, we assume the presence of a cusp, and make hypotheses about the involved variables. In a mechanistic approach, the cusp is derived from basic assumptions or first principles.

The mechanistic approach is much more common in the physical and life sciences. An example is the phase transition in water described by the van der Waals equation. Poston and Stewart (2014) show how the van der Waals equation can be reparametrized to take the form of the cusp equation. The advantage is that we learn how temperature and pressure are related to the control variables of the cusp. This gives us a full understanding of the dynamics of this phase transition.

One model that I will use as a psychological model in chapter 4, section 4.3, is the model of the spruce budworm outbreak, which occurs every thirty to forty years and results in the defoliation of tens of millions of hectares of trees (Ludwig, Jones, and Holling 1978). The model is

$$\frac{dN}{dt} = r_b N \left(1 - \frac{N}{K}\right) - \frac{BN^2}{A^2 + N^2}. \qquad (3.7)$$

Where N is the size of budworm population, r_b is the growth rate, K is the carrying capacity, B is the upper limit of predation, and $1/A$ is the responsiveness of the predator.

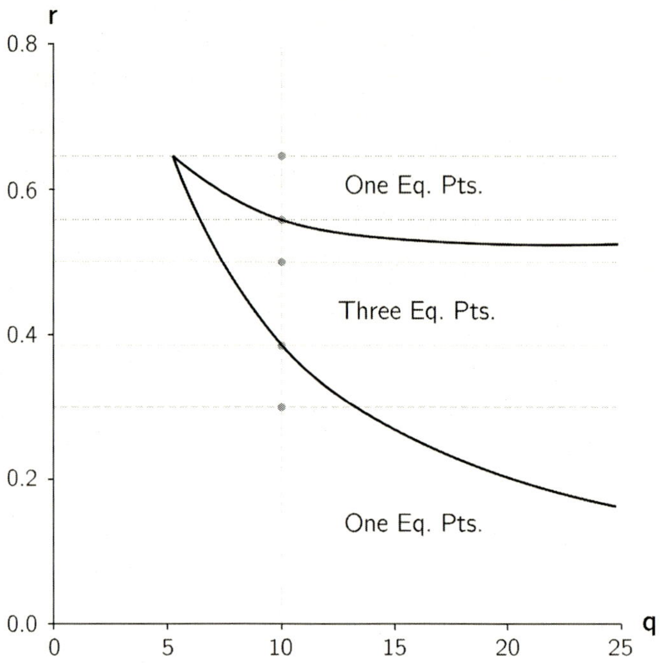

Figure 3.16. The bifurcation diagram of the spruce budworm model. In the bifurcation set, there are two alternative stable states: the normal population level and the outbreak level.

The first part is the logistic growth equation. N will grow to K at a rate r_b. Note that this is a differential equation, not a difference equation. There is no chaos in logistic growth in continuous time. The second part is the predation function and has an increasing shape flattening out at B. The curvature of this function is determined by A. High A makes the function less steep, meaning that predation reacts rather slowly to the increase in budworms (more about the construction of this model later).

The analytical approach to this model is to reparameterize the model so that it takes the form of a cusp. Such reparameterizations are not so easy to do yourself. The idea is to create a smaller set of new variables that are functions of the model parameters. For this model a convenient reparameterization is

$$r = \frac{A r_b}{B} textand\, q = \frac{K}{A}. \tag{3.8}$$

Using these two "constructed" control variables, we can depict the bifurcation lines of the cusp as in figure 3.16.

In later chapters we will discuss psychological examples of a mechanistic approach, but as far as models of transitions are concerned, these are rare. The phenomenological approach is much more common.

3.4.2 PHENOMENOLOGICAL MODELS

The cusp model of attitude is a typical phenomenological model. We simply assume that the cusp is a model of the attitude. Phenomenological models are less convincing than mechanistic models because they do not provide a deep understanding of the underlying mechanisms that drive the system. But in psychology and the social sciences, we cannot be too picky. Compared to many other, verbally stated attitude models, the cusp attitude model is quite precise. It implies a number of phenomena and is testable.

Setting up a phenomenological model is not a trivial task. I suggest some guidelines for this. First, define the behavioral variable. It is important to think about the bistable modes. What are they? What is the inaccessible state in between? Can you have jumps between these states? What is neutral state at the back of the cusp? If you cannot answer these questions, you should reconsider whether a cusp is an appropriate model.

Second, select the control variables. What could be a normal variable and what could be a splitting variable? These are not easy questions. Sometimes there are too many candidates. For the cusp model of attitudes, instead of involvement, we could suggest interest, importance, emotional value, etc. In this case, I think of the splitting axis as a common factor of all these slightly different variables. In other cases, we have no good candidates. In the example in figure 3.14, it is not clear exactly what is being manipulated along the normal axis. If you made a choice, it is good to check whether, at high values of the splitting values, variation of the normal variable may lead to sudden jumps and hysteresis. Also check whether the pitchfork bifurcation makes sense theoretically.[8]

There is another issue here. In some phenomenological models, the control variables are rotated by 45 degrees. The most famous example is Zeeman's (1976) model of dog aggression (fig. 3.17).

Control variables in cusp models can be rotated for ease of interpretation.

[8] Given these guidelines and examples, it is an interesting exercise to develop one's own cusp model, for example, for falling in love. This is a tricky exercise.

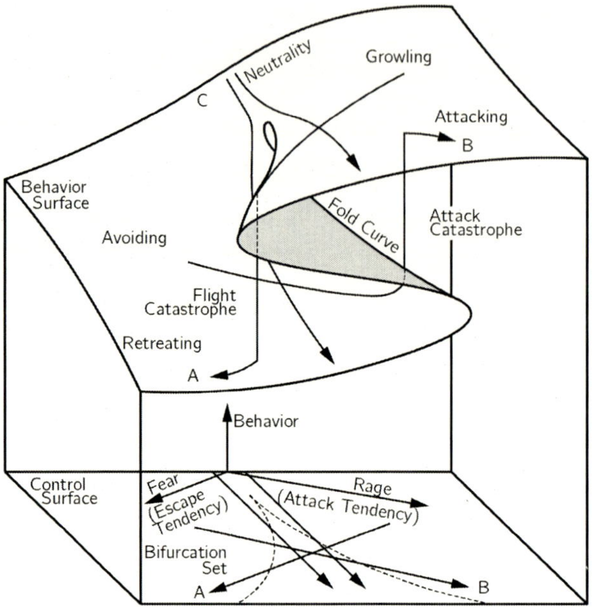

Figure 3.17. Zeeman's dog aggression model with rage and fear as rotated control variables.

The control variables are fear and rage. In such a rotation the normal variable is the difference between fear and rage, while the splitting variable is the sum of fear and rage. Another example can be found in our model of the speed-accuracy trade-off in reaction time tasks (Dutilh et al. 2011). When constructing a phenomenological model, these two options for defining the control variables should be considered.

To explain catastrophic drops in performance in work and sports, Hardy and Parfitt (1991) proposed a cusp model with cognitive anxiety as the splitting factor and physiological arousal as the normal factor. The idea is that at high levels of cognitive anxiety, increases and decreases in arousal lead to sudden changes, including a hysteresis effect. Hardy (1996) presents further tests of this model, which has been criticized by Cohen, Pargman, and Tenenbaum (2003). Extensions to the butterfly model are presented in Guastello (1984) and Hardy, Woodman, and Carrington (2004).

Cusp models have also been developed for addiction (Guastello 1984; Mazanov and Byrne 2006). Witkiewitz et al. (2007) propose using distal risk as the splitting axis and proximal risk as the normal axis. The model is tested

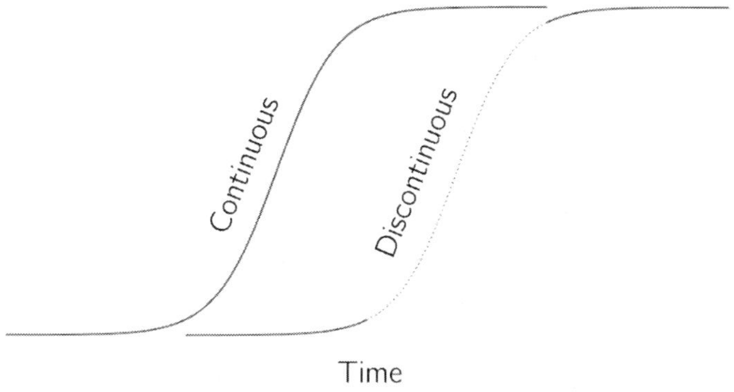

Figure 3.18. Continuous and discontinuous growth curves may look very similar.

using the renowned dataset from Project MATCH, an eight-year, multisite investigation of the effectiveness of various treatments for alcoholism.

As a final example, I mention the model for humor presented by Paulos (2008) in his fascinating book on mathematics and humor. Paulos explains his model in the context of puns. His example is: "Do you consider clubs appropriate for young children?" with the punchline "Only when kindness fails," which is probably only funny to people with children. Paulos uses the rotated control axis as in the dog aggression model. Interpretation of the pun is the behavioral axis. One axis represents the first meaning of "clubs," the other axis represents the second meaning. The bifurcation set represents the ambiguous region. A joke involves a jump from one meaning to another. Paulos claims that this cusp model combines cognitive incongruity theory, various psychological theories of humor, and the release theory of laughter. Tschacher and Haken (2023) propose a related complexity account of humor.

The punch line forces a catastrophic change in interpretation, accompanied by a release of tension through laughter.

3.5 Testing Catastrophe Models

3.5.1 THE CATASTROPHE FLAGS

How sudden is sudden? How can climate changes be seen as transitions between stages (i.e., ice ages) when these transitions take hundreds of years? Even when the ball is rolling toward its new minimum, it takes time to roll. But then what is the difference with an continuous acceleration, such as we see in a logistic growth pattern? The time course of an acceleration and a sudden, discontinuous jump may look very similar (fig. 3.18).

Sudden transitions are not instantaneous, but the in-between states are unstable.

In fact, in terms of time-series data, they may look exactly the same. The main difference is that in the continuous case the intermediate values are stable. An acceleration can be understood as a quadratic minimum that changes its position quickly. If we stop the process by freezing the manipulated control variable in the process, the state will remain at an intermediate value. These intermediate values are all stable values. If we freeze the manipulated variable in a discontinuous process, it will continue to move to a stable state. In this case, the intermediate state is unstable. The ball keeps rolling and unfortunately the climate keeps changing.

In practice using time series data, this is a difficult distinction to make. It means that simple time series are not sufficient to distinguish accelerations from phase transitions. So how do we distinguish between the two processes? In the context of catastrophe theory, Gilmore (1993) proposed the catastrophe flags. These are cusp-related phenomena that can be seen in the data. While no single one of these is sufficient to indicate the cusp, when considered together they provide compelling evidence for its existence.

In the following subsections, I will define the flags and illustrate their applications in psychology using examples. The first flag is the sudden jump.

3.5.1.1 Sudden Jump

The sudden jump is a large fast change in equilibrium behavior.

Although the sudden jump is not sufficient (it could be due to an acceleration), demonstrating a sudden jump in time series is useful (also in relation to other flags). Statistical detection of sudden jumps is possible using a number of techniques. Figure 3.19 presents raw weekly measurements of depressive symptoms using the SCL-90-R depression subscale of a patient who gradually stopped antidepressant medication during the study. The participant and researchers were blind to the dose reduction scheme (Wichers, Groot, and Psychosystems 2016). One question was whether this reduction led to a sudden jump to the depressed state. Using a change point detection method (James and Matteson 2014), we found a jump at 18 weeks with a bootstrapped p-value of .005 (with the null hypothesis of no change point).

Many methods for change point analysis have been developed and compared in Burg and Williams (2022).

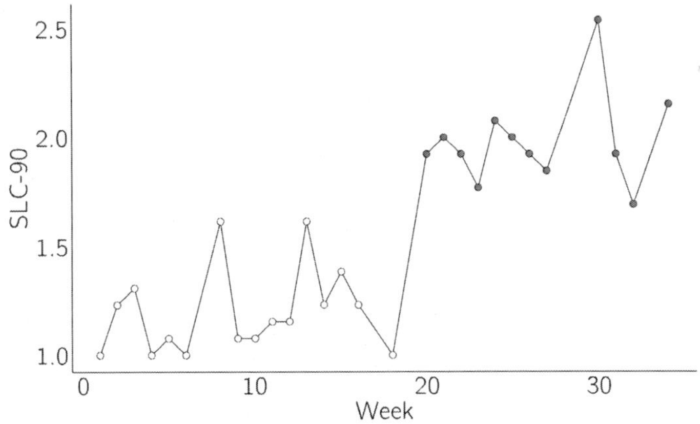

Figure 3.19. A sudden jump to depression (score at the SLC-90) in a patient who gradually quit antidepressant medication during the study.

The code for this figure is:

```
layout(t(1)); par(mar = c(4,4,1,1))
x <- read.table('data/PNAS_patient_data.txt', header = TRUE)
library(ecp) # if error: install.packages('ecp')
e1 <- e.divisive(matrix(x$dep, , 1), sig = .01, min.size = 10)
plot(x$week, x$dep, type = 'b', pch = (e1$cluster-1) * 16 + 1,
     xlab = 'Week', ylab = 'SLC-90', bty = 'n',
     main = 'Jump to depression')
```

3.5.1.2 Multimodality

Multimodality (in the case of the cusp bimodality) is an important and easy-to-use flag, as it can be tested with cross-sectional data. Finite mixture models have been developed to test for multimodality in frequency distributions (McLachlan, Lee, and Rathnayake 2019).

An example is shown in figure 3.20. These data come from a conservation anticipation task, where children have to predict the level of water in the second glass when it is poured over. The resulting data and the fit of a mixture of two normal distributions are shown on the right. The data are clearly bimodal supporting the hypothesis of a transition in conservation learning (van der Maas and Molenaar 1992). These data were used in Dolan and van der Maas (1998) to fit multivariate normal mixture distributions subject to a structural equation model.

The code is:

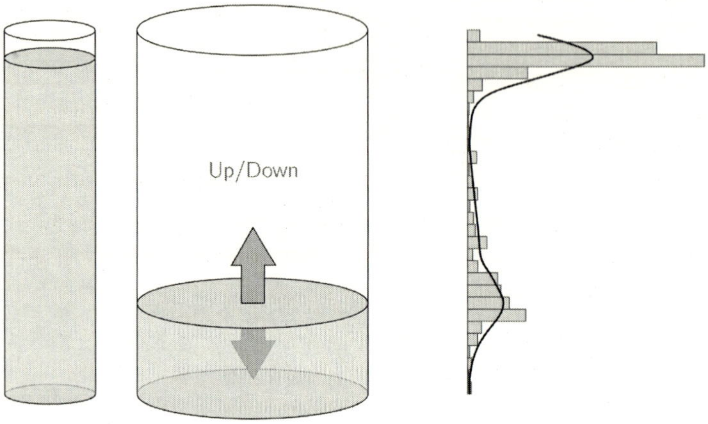

Figure 3.20. Bimodality in the expected heights of water when it is poured into a wider glass. This variation of the Piagetian conservation task is used with children ages five to eight.

```
x <- read.table('data/conservation_anticipation_item3.txt')
x <- unlist(x)
library(mixtools) # if error: install.packages('mixtools')
result <- normalmixEM(x)
plot(result, whichplot = 2, breaks = 30)
```

There is a whole field in statistics focused on multimodality, mixtures, and clustering. There are some blogs that present overviews of the relevant R packages (Arnaud 2021). Several detailed examples from psychology, using hidden Markov models, are presented in Visser and Speekenbrink (2022).

The advantage of multimodality over the sudden jump is that we can test it with cross-sectional data.

To capture a sudden jump in a development process, you need a lot of high-frequency data. Sudden shifts in opinion are also rare. But it is easy to collect data on large numbers of people who are asked to make judgments about statements on an issue such as abortion. If these judgments are bimodally distributed, this is consistent with a phase transition. Bimodal data may also be produced by a process of acceleration, with time series consisting mainly of data values before and after the acceleration. So, bimodality is not sufficient. It can be considered necessary, so I always suggest starting with cross-sectional multimodal studies. If they fail, you might reconsider your hypothesis. I have often looked for multimodality in measures of arithmetic learning and never found anything convincing, which made me rethink my hypothesis.

3.5.1.3 Inaccessibility

Inaccessibility means that certain values of the behavioral variable are unstable. The business card is a good example. Given some vertical pressure, we can try what we want but we cannot force the card to stay in the middle position; it is unstable.

Inaccessibility is relevant to reject the alternative hypothesis that the sudden jump and bimodality are due to an acceleration.

In experiment 2 of Dutilh et al. (2011), we focused on this flag. Our hypothesis was that in simple choice response tasks there is a phase transition between a fast-guessing state and a slower stimulus-driven response state. The idea is that if we force subjects to speed up, there will be a catastrophic decline in performance (from almost 100% correct to 50% correct).

We created a game in which subjects responded to a series of simple choice items (a lexical decision task). The length of the series was not known to the subject. At the end of a series, they were rewarded according to how close their percentage correct was to 75%. Speed was also rewarded, but much less. So, we asked the subject to be in the inaccessible state. The alternative hypothesis, based on information accumulation models, was that there was no phase transition and that responding with 75% accuracy required the correct setting of a boundary (see sec. 4.3).

It appeared that subjects solved the task by switching between the fast-guessing mode and the slower stimulus-controlled mode, even when instructed according to the alternative model. Thus, the 75% intermediate state appeared to be unstable.

3.5.1.4 Divergence

Divergence or the pitchfork bifurcation, the splitting up of an equilibrium, requires the manipulation of the splitting variable. In the case of attitudes, we hypothesize this to be involvement or some related variable. In van der Maas, Kolstein, and van der Pligt (2003), we reanalyzed a dataset from Stouffer et al. (1949), which Latané and Nowak (1994) presented as evidence for the cusp model. The attitude concerned demobilization (from 0, unfavorable, to 6, favorable), and respondents were asked to indicate how strongly they felt about their answer (from intensity 0 to intensity 5). For low intensities of feeling, the data are normally distributed whereas for higher intensities, data are bimodally distributed (see fig. 3.21).

After testing for multimodality, testing for divergence is a sensible next step.

Figure 3.21. The pitchfork bifurcation in attitudes. The dotted lines represent the fit of the cusp model to these data. This technique will be discussed in section 3.5.

3.5.1.5 Hysteresis

Hysteresis, the lagging behind of the sudden jump, requires sophisticated manipulation of the normal control variable.

To test for hysteresis, we need to slowly increase and decrease the normal variable and test whether sudden jumps occur with a delay. We have demonstrated hysteresis in proportional reasoning using Piaget's balance scale test in which a specific dimension (distance from the fulcrum) was systematically varied (Jansen and van der Maas 2001). We also hypothesized that speeding up subjects in response time tasks would eventually lead to a catastrophe in accuracy. To support this claim, we demonstrated bimodality in response times and hysteresis in the speed-accuracy trade-off (Dutilh et al. 2011). To support the cusp model of multistable perception, we used the quartet motion paradigm (Ploeger, van der Maas, and Hartelman 2002). In this perceptual paradigm two lights are presented simultaneously, first a pair from two of the diagonally opposite corners of the rectangle, and then a second pair from the other two diagonally opposite corners of the rectangle. Usually, either vertical or horizontal apparent motion is perceived. By gradually increasing or decreasing the aspect ratio (i.e., the ratio of height to width of the quartet), hysteresis in the jumps between the two percepts was demonstrated (see fig. 3.22).

In Ploeger, van der Maas, and Hartelman (2002), we used a special design, the method of modified limits, to rule out the alternative explanation that hysteresis is simply due to delayed responses. It could be that the switches always occur in the middle (at an aspect ratio of 1), but the self-report is delayed. In the modified limits method, subjects do not respond during a trial, only after the entire trial. By varying the length of the trials,

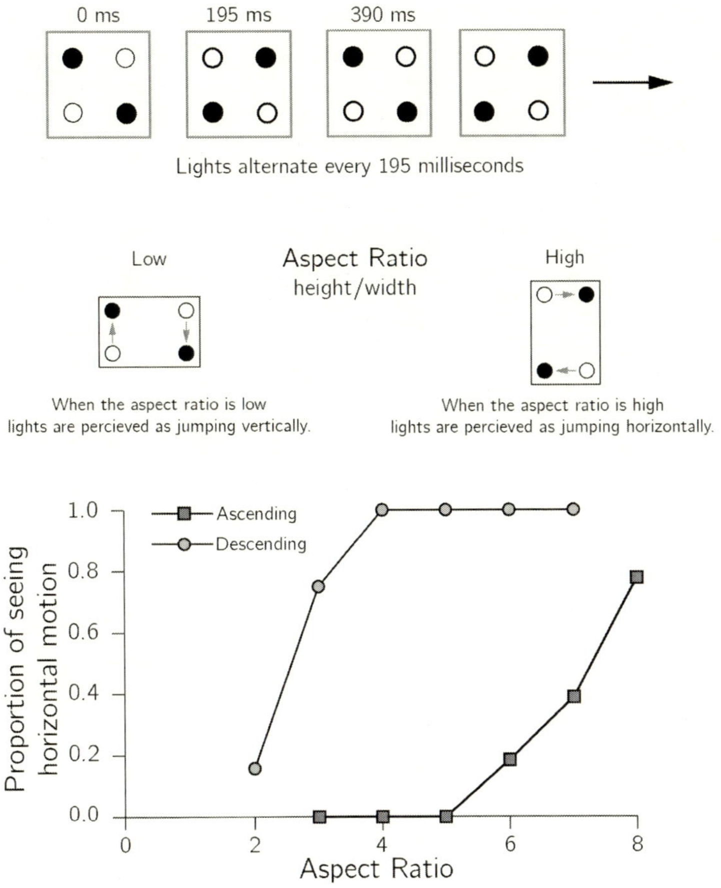

Figure 3.22. Hysteresis in the perception of apparent motion. Switches between the perception of vertical or horizontal apparent motion occur when the aspect ratio (horizontal axis) is varied. The aspect ratio is the ratio of height to width of the quartet. *(Adapted from Ploeger, van der Maas, and Hartelman (2002) with permission.)*

it is possible to determine at which parameter value the subject perceives a switch.

3.5.1.6 Anomalous Variance, Divergence of Linear Response, and Critical Slowing Down

Gilmore's last three flags—anomalous variance, divergence of linear response, and critical slowing down—are indicators that occur near the bifurcation lines. They are also known as early warning signals and are a popular topic of research (Dakos et al. 2012).

Early warnings are indicators or signals that precede and predict transitions within a system, allowing for anticipation and potentially preventative action.

Anomalous variance occurs because near a bifurcation point the second derivative diminishes, meaning that the minimum becomes less deep. Assuming there is always some perturbation of the state, this will lead to larger fluctuations in the state.

Divergence of linear response is the size of the effect of a small perturbation of the state, which will be greater near a bifurcation point. It will also take longer to return to equilibrium. This delay in return time is known as critical slowing down and is also studied in other approaches to nonlinear dynamical systems (e.g., synergetics, Haken 1977). Examples of applications in psychology can be found in Leemput et al. (2014) and Olthof et al. (2020). A somewhat critical review is provided in Dablander et al. (2023).

In my experience, the problem with early warning signals is that both type 1 and type 2 errors should be low for predicting transitions. This is challenging even in simulations, let alone in noisy psychological data. It would be fantastic if these early warnings really worked. For example, being able to predict a relapse into depression or addiction would be of great clinical value.

Together, the catastrophe flags provide a methodology for phase transition research in psychology. A single flag may not be sufficient, but the combination is. For example, the combination of evidence for inaccessibility and hysteresis is convincing. I have given psychological examples of most of the flags. Which flags to use in a particular case depends on the knowledge and experimental control of the control variables. Another approach is to fit the cusp model directly to the data. This is the subject of the next section.

3.5.2 FITTING THE CUSP TO CROSS-SECTIONAL DATA

3.5.2.1 Cobb's Maximum Likelihood Approach

In a series of papers, Loren Cobb and colleagues (Cobb and Zacks 1985; Cobb 1978) developed a maximum likelihood approach[9] to fit the cusp catastrophe to data consisting of cross-sectional measurements of X, a, and b. We have implemented this approach in a cusp R package described in Grasman, van der Maas, and Wagenmakers (2009).

[9]The is method finds the parameter values that make the observed data most probable.

The basic idea is to make catastrophe theory, a deterministic theory, stochastic by adding a stochastic term, called Wiener noise (with variance σ^2), to equation 3.2[10]:

$$dX = -V'(X)dt + \sigma dW(t). \tag{3.9}$$

It is important to note that this type of stochasticity is not the same as measurement noise. Measurement noise—that is, ε in $Y = X + \varepsilon$—does not affect the dynamics of X. Wiener noise does; it is part of the updating equation of X itself. This stochastic differential equation is associated with a probability distribution of the form:

A stochastic differential equation (SDE) is a differential equation that incorporates a term representing random fluctuations.

$$f(X) = \frac{1}{Z\sigma^2} e^{\frac{-V(X)}{\sigma^2}}, \tag{3.10}$$

where Z is a normalizing constant[11] necessary to ensure that the area under $f(X)$ is 1. This may look complicated, but for the quadratic case $V(X) = \frac{1}{2}X^2$, this results in the standard normal distribution, with $Z = \sqrt{2\pi}/\sigma$.

As in the case of the normal distribution, we want to allow for some transformations of the variables. To simplify the necessary statistical notation, we write the cusp as $V(y) = -\alpha y - \frac{1}{2}\beta y^2 + \frac{1}{4}y^4$. The probability distribution for the cusp is:

$$f(y) = \frac{1}{Z\sigma^2} e^{\frac{\alpha y + \frac{1}{2}\beta y^2 - \frac{1}{4}y^4}{\sigma^2}}. \tag{3.11}$$

As in regression models, the cusp variables are modeled as linear function of measured variables. That is, the dependent variables $Y_{i1}, Y_{i2}, \ldots, Y_{ip}$ and the independent variables $X_{i1}, X_{i2}, \ldots, X_{iq}$, for subjects $i = 1, \ldots, n$, are related to the cusp variables as follows:

$$\begin{aligned} y_i &= w_0 + w_1 Y_{i1} + w_2 Y_{i2} + \ldots + w_p Y_{ip}, \\ \alpha_i &= a_0 + a_1 X_{i1} + a_2 X_{i2} + \ldots + a_q X_{iq}, \\ \beta_i &= b_0 + b_1 X_{i1} + b_2 X_{i2} + \ldots + b_q X_{iq}. \end{aligned} \tag{3.12}$$

[10] Many different notations exist for this. Perhaps clearer is $dX(t) = -V'(X(t))\,dt + \sigma dW(t)$, as both dX and dW depend on time.

[11] For consistency with later chapters, I define Z differently from the notation in Grasman, van der Maas, and Wagenmakers (2009). It is the inverse of Z in that paper.

My number-one rule when using statistical techniques: Never use a statistical technique on real data before you have tested it on simulated data.

By estimating these regression parameters, we fit the cusp model to empirical data. The cusp package in R makes this possible. I will first demonstrate this using simulated data. I strongly recommend this approach. First, it forces you to understand what the statistical technique actually does, and second, it gives you a way to test the power and investigate violations of the technique's assumptions.

```
library(cusp) # if error: install.packages('cusp')
set.seed(1)
X1 <- runif(1000) # independent variable 1
X2 <- runif(1000) # independent variable 2
# to be estimated parameters:
w0 <- 2; w1 <- 4; a0 <- -2; a1 <- 3; b0 <- -2; b1 <- 4
# sample Y1 according to cusp using rcusp and the parameter values:
Y1 <- -w0/w1 + (1/w1) * Vectorize(rcusp)(1, a0+a1*X1, b0+b1*X2)
# collect 'measured' variables in data:
data <- data.frame(X1, X2, Y1)
```

I recommend doing some descriptive analysis first. With `hist(data$Y1)` we can inspect whether there is some indication of bimodality. $X2$ is the splitting variable, so perhaps we see stronger bimodality with `hist(data$Y1[data$X2>mean(data$X2)]`). The function pairs in R, `pairs(data)` , is also always recommended. In this perfect simulated case, you will already see strong indications of the cusp. Now we fit the full model with α and β both as function of $X1$ and $X2$.

```
fit <- cusp(y ~ Y1, alpha ~ X1+X2, beta ~ X1+X2, data)
summary(fit)
```

Table 3.1 provides a summary.

Note that we fit a model with too many parameters. We also estimated a_2 and b_1 (because the model was specified as alpha ~ X1+X2, beta ~ X1+ X2). These estimates are not significantly different from 0. The other parameters are estimated reasonably close to their true values, since the true values fall within the confidence interval of the estimates (defined by twice the standard error on either side). We expect a better fit in terms of AIC and BIC when we fit a reduced model without a_2 and b_1. These fit indices penalize the goodness of fit (e.g., the log-likelihood) for the number of parameters used to discourage overfitting and to promote model parsimony.

Table 3.1. The parameter estimates including standard errors and p-values generated by the cusp package.

Coefficients	Estimate	Std. Error	Z-Value	PR(>|z|)
a[Intercept]	−2.13	0.19	−11.0	< 2e−16***
a[X1]	3.11	0.22	14.2	< 2e−16***
a[X2]	0.15	0.17	0.9	0.39
b[Intercept]	−2.29	0.34	−6.7	2.66e−11***
b[X1]	−0.09	0.33	−0.3	0.79
b[X2]	4.40	0.27	16.5	< 2e−16***
w[Intercept]	1.98	0.07	27.6	< 2e−16***
w[Y1]	3.97	0.10	38.0	< 2e−16***

```
fit_correct_model <- cusp(y ~ Y1, alpha ~ X1, beta ~ X2, data)
summary(fit_correct_model)
```

Table 3.2. The comparative fit measures AIC, AICc, and BIC indicate that the reduced model should be the model of choice.

	R.Squared	logLik	npar	AIC	AICc	BIC
Full model	0.428	−1058.7	8	2133.3	2133.5	2172.6
Reduced model	0.426	−1059.0	6	2130.0	2130.1	2159.5

The next simulation demonstrates that we can detect hysteresis using this approach. We simulate data with $-2 < \alpha < 2$, and fixed β. If $\beta < 0$ we have no hysteresis, but if $\beta > 0$, we do have hysteresis. With the code below we simulate datasets for different β and compare the goodness of fit between the linear and cusp model. Figure 3.23 summarizes the results. Note that a lower BIC indicated the better-fitting model.

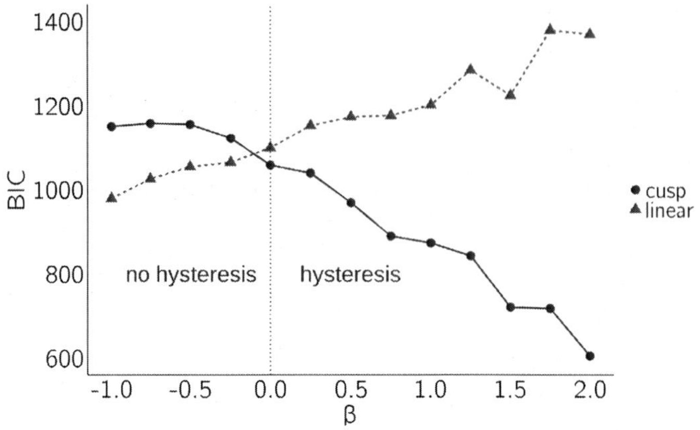

Figure 3.23. At the back of the cusp (low b_0), the cusp is approximately linear, the BIC favors this simpler model (dotted line) over the cusp mode (solid line).

```
set.seed(10)
n <- 500
X1 <- seq(-1, 1, le = n) # independent variable 1
a0 <- 0; a1 <- 2; b0 <- 2 # to be estimated parameters
b0s <- seq(-1, 2, by = .25)
i <- 0
dat <- matrix(0, length(b0s), 7)
for (b0 in b0s){
  i <- i + 1
  Y1 <- Vectorize(rcusp)(1, a1 * X1, b0)
  data <- data.frame(X1, Y1) # collect 'measured' variables
  fit <- cusp(y ~ Y1, alpha ~ X1, beta ~ 1, data)
  sf <- summary(fit)
  dat[i, ] <- c(b0,
              sf$r2lin.r.squared[1], sf$r2cusp.r.squared[1],
              sf$r2lin.bic[1], sf$r2cusp.bic[1],
              sf$r2lin.aic[1], sf$r2cusp.aic[1])
}
par(mar = c(4,5,1,1))
matplot(dat[,1], dat[,4:5], ylab = 'Bic', xlab = 'b0',
        bty = 'n', type = 'b', pch = 1:2, cex.lab = 1.5)
legend('right', legend = c('linear','cusp'), lty = 1:2,
       pch = 1:2, col = 1:2, cex = 1.5)
abline(v = 0, lty = 3)
text(-.5, 800, 'no hysteresis', cex = 1.5)
text(.5, 800, 'hysteresis', cex = 1.5)
```

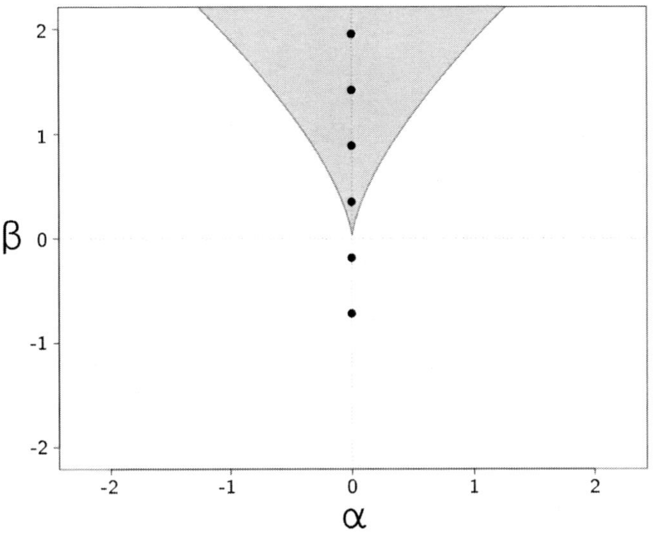

Figure 3.24. Placement of the data of figure 3.21 in the bifurcation set.

3.5.2.2 Empirical Examples

In Grasman, van der Maas, and Wagenmakers (2009), we present several examples with real data. As another example, we use Stoufer's data, which we used as an example of divergence before (see fig. 3.21).

```
x <- read.table('data/stoufer.txt')
colnames(x) <- c('IntensityofFeeling', 'Attitude')
fit <- cusp(y ~ Attitude,
        alpha ~ IntensityofFeeling,
        beta ~ IntensityofFeeling, x)
summary(fit)
```

Inspection of the parameter estimates shows that, as expected, intensity of feeling only loads on the splitting axis and not on the normal axis. Figure 3.24 shows the location of the data in the bifurcation set (`plot(fit)`).

Another example is the conservation dataset of Bentler (1970), which contains the scores on a twelve-item test from a conservation test of 560 children from eight different age groups (fig. 3.25). These data are expected to be bimodal and to move along the normal axis (van der Maas and Molenaar 1992).

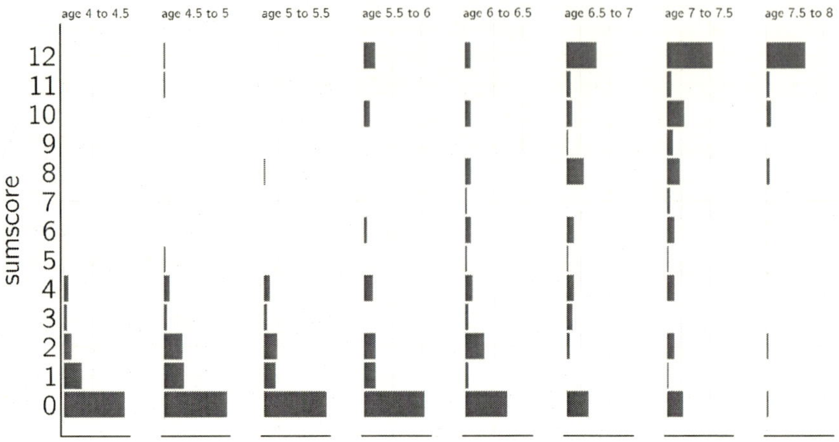

Figure 3.25. In the Bentler conservation data, the sumscore distribution is bimodal, and the weights of the modes shift with age.

```
x <- read.table('data/bentler.txt', header = TRUE)
layout(t(1:8))
age <- c('age 4 to 4.5', 'age 4.5 to 5', 'age 5 to 5.5',
         'age 5.5 to 6', 'age 6 to 6.5', 'age 6.5 to 7',
         'age 7 to 7.5', 'age 7.5 to 8')
for(i in 1:8){
  if(i == 1) {par(mar = c(4,3,2,1)); names = 0:12} else
    {names = ''; par(mar = c(4,1,2,1))}
  barplot(table(factor(x[x[,1] == i,2], levels = 0:12)),
          horiz = TRUE, axes = FALSE,
          main = age[i], xlab = '',
          names = names, cex.main = 1.5, cex.names = 1.5)
}
fit <- cusp(y ~ score, alpha ~ age_range, beta ~ age_range, x)
summary(fit)
plot(fit)
```

This is supported by results of the cusp fit. You can verify that a model with beta ~ 1 fits better according to the AIC and BIC.

A great exercise we have often used in the classroom is to build a Zeeman machine, collect data with it, and fit the cusp model to the data (see Grasman, van der Maas, and Wagenmakers 2009 for details). Zeeman invented this machine to demonstrate the properties of the cusp. Our students were rewarded for the quality of the model and the artistic value of their Zeeman machine (fig. 3.26).

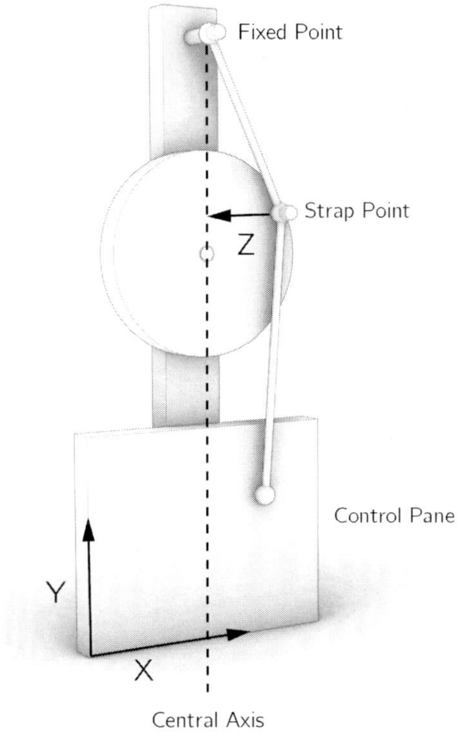

Figure 3.26. Zeeman's catastrophe machine. It consists of a rotating disk and two elastic bands. The first elastic band is attached to a fixed point and the strap point. The end of the other elastic (red dot) is moved by hand through the control plan. The strap point moves according to the cusp catastrophe. Data is gathered by collecting a set of X, Y, and Z values. Typically, 50 to 100 data points are sufficient to apply the cusp fit function in R.

3.5.2.3 Evaluation

A few final remarks: First, Cobb's method can be used with cross-sectional data. Data points should be independent. To test for hysteresis in time series, other approaches are required. One option is to use hidden Markov models as in Dutilh et al. (2011).

Cobb's method is not valid for time series.

Second, there are some issues with Cobb's approach that are due to fundamental differences between probability distributions and potential functions. The latter can be transformed in many ways (so-called local diffeomorphisms) without changing the qualitative properties of the cusp. With the added constraint on probability distributions (area = 1), the same transformations can lead to qualitative effects, such as a change in the number of modes. Wagenmakers et al. (2005) suggest a solution to this problem for time series.

Third, two alternative approaches have been proposed. Both Guastello's (1982) change score least square regression approach and the Gemcat approach (1987) use the first derivative of the cusp as point of departure. A problem with both approaches is that they do not distinguish between stable and unstable equilibrium states. Data points in the inaccessible region improve the fit of the model, whereas they should decrease the fit. Alexander et al. (1992) provide a detailed critique.

3.6 Criticism of Catastrophe Theory

Rosser (2007) speaks of the rise and the fall of catastrophe theory. The hype following the publication of Zeeman (1976) in *Scientific American*,[12] in which he introduced the phenomenological application of catastrophe theory in the behavioral and social sciences, led to a strongly worded reply by Zahler and Sussmann (1977) in *Nature*.

Because people still refer to this paper when we use catastrophe theory in our work, I will briefly respond to Zahler and Sussmann's main points of criticism. In their introduction, they state that there may be legitimate uses of catastrophe theory in physics and engineering.

They do not question the correctness or importance of catastrophe theory as a purely mathematical subject.

They raise ten points, some of which I have already addressed. For example, their first point is about how sudden a jump actually is, but they call this a less serious criticism. As I explained earlier, it is not the suddenness that matters but whether or not the intermediate states are unstable.

A number of points are about inferring a cusp from data, which was indeed done rather superficially in Zeeman's earlier work. They point out that there are no testable predictions, that the location of the cusp can be shifted, and that there is no way to decide whether the data fit the cusp. I hope to have shown that these problems are largely solved: the catastrophe flags allow us to make new testable predictions, and with Cobb's maximum likelihood approach we can fit the model as we would do with any other statistical model in modern science. Of course, one can be critical of the

[12] To see Zeeman at work, I recommend the BBC documentary *Case Study Catastrophe Theory Maths Foundation Course* (https://www.youtube.com/watch?v=myDvcvox1V4&t=1435s).

use of statistics in psychology and the social sciences, but these criticisms are not specific to catastrophe theory.

Another, somewhat inconsistent, line of criticism is that many catastrophe models in psychology and the social sciences are just wrong and inconsistent with the data (which could be true), while also not falsifiable. But you cannot have it both ways: if it is wrong or inconsistent with the data, it is falsifiable. Nevertheless, I agree that it is important to think about falsifiability. Theories in psychology tend to be moving targets. As soon as someone finds an empirical result that contradicts the theory, the theory is quickly modified.

Then Zahler and Sussmann point out that catastrophe theorists often try to make a discrete variable into a continuous one. Their example is aggression, which they believe is inherently discrete. They call Zeeman's interpretation of aggression as a continuous family of behaviors absurd and utterly meaningless. This may be a bit strong. We can think of situations in which aggression can vary from mild to severe, or from verbal to physical, directed at a person's belongings, mild physical directed at the person, to severe physical. A rich ordering of aggressive acts is very useful for describing domestic violence. Sometimes the change along these acts or variants is gradual, and other times sudden. Whether such an ordering can be treated as a quantitative continuum is one of the most difficult questions in our field (Borsboom et al. 2016; Michell 2008).

Zahler and Sussmann's final point is that there are better alternatives, such as quantum mechanics, discrete mathematics, and bifurcation theory. There is work on quantum mechanics in psychology (especially in the context of consciousness), but whether it will lead to breakthroughs in this field remains to be seen. Discrete mathematics may be an alternative in some cases (e.g., to model symbolic thinking). I see catastrophe theory as a special branch of bifurcation theory, especially useful when the system under study is difficult to describe in terms of mathematical equations. This goes back to the distinction between phenomenological and mechanistic models. I think we should put more effort into developing mechanistic models based on first principles. More on this in the next chapters.

The empirical program, using catastrophe flags in conjunction with Cobb's method for fitting cusp models, bypasses much of the previous criticism of catastrophe theory.

Loehle (1989) presents an excellent discussion on the usefulness of catastrophe theory in the context of modeling ecosystems. He concludes that "an unresolved problem in applying catastrophe models is that of testing the goodness of fit of the model to data," but this problem has now been largely solved.

3.7 Conclusion

Psychologists are often concerned with psychological types and classes, stages and phases, and the transitions between them. Our thinking about transitions becomes much clearer and more advanced when we know the basics of bifurcations.

Catastrophe theory comes with a toolbox for the behavioral and social sciences. We can build phenomenological models, test for catastrophe flags, and even fit cusp models to data. With the development of this toolbox, most of the criticisms of catastrophe theory lose their relevance.

However, there is room for improvement. Phenomenological models have limited explanatory power. As explained in the next chapter on dynamical system models, it is possible to create more mechanistic models that support the use of phenomenological models. In chapter 6, section 6.3, another option is introduced using networks. I will demonstrate that the behavior of the Ising network model for attitudes is governed by the cusp model, which is very similar to the cusp model proposed for the attitude toward abortion.

3.8 Exercises

1. The equilibria of the fold are $X = \pm\sqrt{\frac{a}{3}}$. This can be checked by setting the first derivative to 0. Show this. (*)

2. In Zeeman's dog aggression model, fear and rage are "rotated" control variables. How can we translate this to a model with unrotated axes? Provide the equations that specify the normal and splitting axis as function of fear and rage. (*)

3. Derive the equation for the bifurcation lines of the cusp ($27a^2 = 4b^3$), by setting the first and second derivatives to 0. Plot the bifurcation lines in GeoGebra or Desmos. (**)

4. Some insight into the butterfly catastrophe $V(X) = -aX - bX^2 - cX^3 - dX^4 + X^6$ can be gained by entering the equation in free online graphing calculators such as Desmos or GeoGebra. Set a, b, c, d to

0, −5, 0, 5. Then start varying a and c. What is the difference in the effect of these two parameters on the appearance and disappearance of attractors? (**)

5. Set up a phenomenological cusp for falling in love. Follow my guidelines (see sec. 3.4). (**)

6. Check whether indeed the Bentler data fit better when `age_range` only loads on the normal axis (according to the AIC and BIC). What is the correct specification of `beta` in `cusp()` in this case? (*)

7. What is the best fitting cusp model (according to the BIC) for this tricky dataset created with this R code? Why? (**)

```
n <- 500
# sample z :
z <- Vectorize(rcusp)(1, .7*rnorm(n), 2+2*rnorm(n))
x <- rnorm(n)
y <- rnorm(n)
data <- data.frame(z, x, y) # collect variables in data
```

8. Build a Zeeman machine, collect data, and fit the cusp (see ex. III of Grasman, van der Maas, and Wagenmakers 2009). What is your best fitting model? Provide a plot of the data in the bifurcation set and a picture of your Zeeman machine. (**)

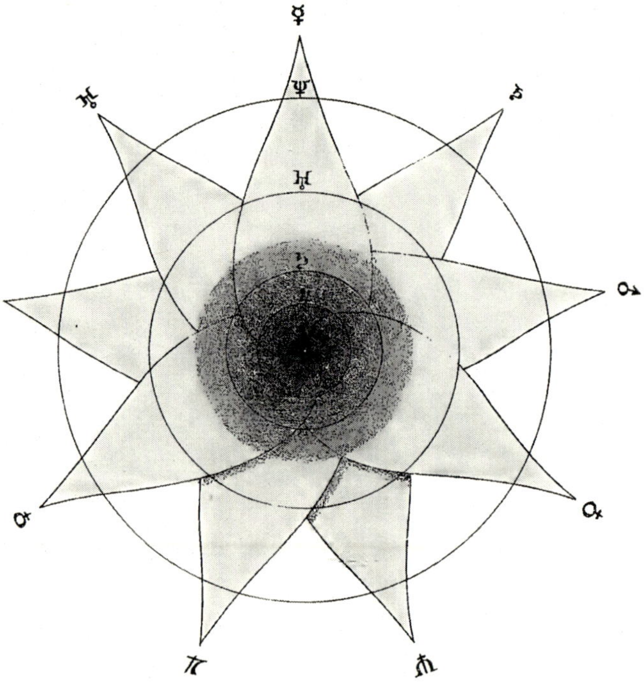

4

BUILDING DYNAMIC SYSTEM MODELS

4.1 Introduction

Suppose you are in a bar in Amsterdam and someone asks if you would like another beer. The number of drinks you have already had will probably influence your decision. Perhaps your self-control, whatever it may be, kicks in and you refuse, even though the alcohol already in your system may be interfering with that self-control. Or you may have reached your limit and simply collapse. In this chapter, we will see how such a decision-making process can be modeled using nonlinear differential equations.

Th… …ersion of this… …ng code block… …an be view… …d at: *santaf… …ub.io/ Comp…*

This form of modeling is often called nonlinear dynamical systems theory (NLDST), another branch of the complex-systems approach. We saw examples of nonlinear dynamical system models in earlier chapters. The logistic map is an example of a discrete-time nonlinear dynamical model defined as a difference equation. The catastrophe models are also dynamical systems governed by a potential function. In chapter 3, section 3.4, I made a distinction between phenomenological modeling (assuming the cusp) and mechanistic modeling (deriving the cusp from first principles). Here we will focus on the more mechanistic construction of dynamical system models.

A nonlinea… system is on… the change of syst… …bles over time is gove… by nonlinear equations, resulting in complex behavior such as chaos and bifurcations.

In psychology, following the principle of parsimony (Occam's razor), we must start with simple models. We don't have many first principles to start with, and our data are often limited, making model testing difficult. But we can learn a lot from other disciplines. Nonlinear dynamical systems have been developed in all the natural sciences, but my main inspiration comes from mathematical biology, especially ecological modeling (Murray 1989). Mathematical psychology is generally less developed than mathematical biology, but this depends somewhat on the subfield. In areas such as neural modeling, speeded decision-making, memory, choice, and psychometrics, there are advanced models, and I will provide some examples later in this chapter.

I will first present a basic overview of dynamical systems modeling in other sciences. Then I will discuss applications in psychology. I refer to more advanced sources when necessary. I recommend the book by Gottman et al. (2002) for its clear and basic explanation of the mathematical aspects of dynamical systems modeling. Strogatz's online lectures on nonlinear dynamics and chaos and his book (2018) are very helpful.[1] Murray's book (2002) on mathematical biology is also highly recommended. Meadows's *Thinking in Systems* (2008) offers a basic introduction.

This chapter will be hands-on again. We will use the Grind package in R to simulate dynamical systems models. Grind (de Boer 2018) is based on the R packages deSolve and rootSolve (Soetaert, Petzoldt, and Setzer 2010). It facilitates numerical integration, phase plane analysis, and stability analysis of steady states.[2] At the end of the chapter, I will introduce causal-loop diagrams and an open-source tool, Insight Maker, that makes it easy to create causal-loop diagrams. We will also use Insight Maker to simulate dynamical systems models, for which I will provide some examples.

4.2 Basic Concepts

4.1.1 BACK TO THE LOGISTIC EQUATION

We saw the logistic equation in the form of the logistic map (sec. 2.2), where time progressed in discrete steps. The logistic map is a difference equation, $X_{t+1} = f(X_t)$, but in this chapter we will focus ondifferential equations in continuous time. We will limit ourselves to ordinary differential equations (ODEs). The ODE for logistic growth[3] is:

In ordinary differential equations, we take the derivatives with respect to only one variable.

$$\frac{dX}{dt} = rX(1 - X). \tag{4.1}$$

The change in X is a function of X itself. The exponential growth term rX dominates when X is close to 0, but the growth levels off as X approaches 1. A solution to this equation expresses X_t as a function of the

[1] https://www.youtube.com/playlist?list=PLbN57C5Zdl6j_qJA-pARJnKsmROzPnO9V

[2] The manual can be found at https://github.com/hansschepers/grindr/blob/master/inst/documentation/GRIND%20tutorial.pdf.

[3] In many texts $\frac{dX}{dt}$ is written as $\dot{X}$.

initial state X_0. In simple cases we can do this using the math we learned in high school. For exponential growth $dX/dt = rX$, this is the derivation:

$\frac{dX}{dt} = rX$	
$\frac{dX}{X} = rdt$	by separation of variables
$\int \frac{dX}{X} = \int rdt$	integrate
$\ln X = rt + C$	assuming $X \geq 0$
$X = e^{rt+C} = e^C e^{rt}$	by taking the exponent
$X_0 = e^C e^{r0} \Longrightarrow X_0 = e^C$	compute the integration constant

$$\Longrightarrow X_t = X_0 e^{rt}. \tag{4.2}$$

But for more complex models, such an analytical solution is out of scope and numerical solutions (by simulation) are required. This is not the preferred choice. These simulations can be slow, they may accumulate rounding errors, and it can be difficult to search the entire parameter space, especially when multiple parameters are involved.

The naive implementation of differential equations in R is risky. This would involve a for loop:

```
x <- x0 <- .1 # initial value
r <- .5 # growth rate
dt <- .00001 # time step in simulation
t <- 10  # Nt, time we want to know the value of x
timesteps <- t/dt # required time steps given t  and dt
for(i in 2:timesteps) # note the 2 to use the starting value
  x[i] <- x[i-1] + r * x[i-1] * dt
x0 * exp(r * timesteps * dt) # analytical solution
x[timesteps] # compare
timesteps # length of simulation
```

where dt must be chosen by hand. If you test some values of dt, you will see that a value too high (.5) leads to a solution (`x[timesteps]`) that is different from the analytical solution. But if we set dt very low (.00001), it takes unnecessarily long.

Solvers are specialized algorithms designed to numerically approximate solutions to ODEs and handle the complexities of integrating these equations over time to predict the evolution of system states under different initial conditions and parameters.

This is why we use solvers, numerical methods for ordinary differential equations. We will use the R package Grind, although many other methods are available in R. One could also directly use the R packages deSolve and rootSolve by Soetaert, Petzoldt, and Setzer (2010), on which the Grind package is based. Grind has to be installed from GitHub using:

```
install.packages("remotes")
remotes::install_github("hansschepers/grindr")
```

The packages required are:

```
library(deSolve)
library(rootSolve)
library(FME)
library(Grind)
```

The code consists of defining the model, the parameters p, and the initial values s. Main functions are `run()`, `plane()`, `newton()`, `continue()`, and `fit()`. They will be introduced using examples. With `run()` we generate a time series for the model.

```
model <- function(t, state, parms) {
  with(as.list(c(state, parms)), {
    dX <- r * X # the exponential model
    return(list(c(dX)))
  })
}
p <- c(r = .5) # parameter r
s <- c(X = .1) # initial value
run(tmax = 5) # run until t = 5, numerical solution
s['X'] * exp(p['r'] * 5) # compare with analytical solution
```

We don't have to worry about time steps anymore, and the numerical and analytical solutions converge. This is of course a trivial use of an ODE solver, but much more can be done.

In analyzing the behavior of a dynamical system, first we want to know what the equilibria X^* are. To do this, we need to set the time derivative equal to 0, $dX/dt = 0$. For the exponential function, this is simply $rX = 0$, that is, when X is 0. Second, we want to determine whether these equilibria are stable or unstable. Whether $X^* = 0$ is stable can be determined by checking the second derivative in X^*. If this derivative is larger than 0, then the fixed point is stable. The second derivative is $-r$, as $dX/dt = -V'(X)$, so $X^* = 0$ is an unstable fixed point whenever $r > 0$ and stable whenever $r < 0$. You can check this in Grind by using r values of $-.1$ and $.1$, and start values equal to or just above or below 0.

For equation 4.1, the logistic function, we also want to know the equilibria, the stable and unstable fixed points. To do so we follow the same steps as for the exponential function (see exercises).

The continuous-time implementation of the logistic function is somewhat boring compared to its discrete-time variant that we studied in section 2.2. The difference is that the overshooting and undershooting do not occur in continuous time. By changing the logistic model in Grind to:[4]

```
dX <- r * X * (1-X) - X
```

and using `method = 'euler'` in the `run()` function, you can simulate the discrete-time logistic map. Check if you get chaos for $r = 4$. Use the Euler method only in special cases, as it is generally the least accurate approach.

4.1.2 THE LOTKA–VOLTERRA MODELS

Perhaps the best-known population models are the Lotka–Volterra equations (Murray 2002). These consist of coupled differential equations, one for the density of the prey and one for the density of the predator:

The Lotka–Volterra models describe the dynamics of biological systems in which two species interact, one as a predator and the other as prey.

$$\begin{aligned} \frac{dN}{dt} &= aN - bPN, \\ \frac{dP}{dt} &= cPN - dP, \end{aligned} \tag{4.3}$$

where N and P refer to the sizes of the prey and predator populations, a and c determine the growth rates, and b and d control the mortality rates. Note that the mortality rate of prey depends on both N and P, while the mortality rate of predators depends only on P. Similarly, the growth terms are also asymmetric, predators increase as a function of both N and P, as they eat prey. We will follow the simple example provided by Wikipedia (on Lotka–Volterra equations).

To implement this model in Grind, we use:

[4] The $-X$ is added because the difference equation has the form $X_{t+1} = f(X_t)$, so the change dX is thus $f(X_t) - X_t$.

```
LV <- function(t, state, parms) {
  with(as.list(c(state, parms)), {
    dN <- a * N - b * P * N
    dP <- c * P * N - d * P
    return(list(c(dN, dP)))
  })
}
p <- c(a = 1.1, b = .4, c = .1, d = 0.4) # parameters
s <- c(N = 10, P = 10) # 10 baboons and 10 cheetahs
```

Some typical uses of Grind are:

```
layout(1:2)
# set tstep to low value :
data <- run(odes = LV, tstep = .01, table = TRUE)
# phase plot for different starting values:
plane(odes = LV, portrait = TRUE,
      ymax = 17, xmax = 50, tstep = 0.1, grid = 4)
```

The plane function makes a phase plot with N and P as axes. The black points are initial states. What we learn from this is that the equilibrium of the Lotka–Volterra equations is a limit cycle that depends on the choice of the initial conditions.

A well-known improvement to this model is to make the prey growth density dependent by using the logistic equation. This can be done by setting `dN <- a*N *(1-N) - b*P*N` in the model. This is the case used as an example in the Grind tutorial, which I highly recommend reading (de Boer 2018). It also contains the appropriate parameter values for this model variant. In this density-dependent model, there are fixed points, in contrast to the original model. This shows that such model choices can have a large effect.

The Susceptible–Infected–Recovered (SIR) model is a basic epidemiological model that divides a population into susceptible (S), infected (I), and recovered (R) individuals.

A famous example of a system of three coupled differential equations is the Susceptible–Infected–Recovered (SIR) model used to model infectious diseases and to understand the impact of interventions on disease dynamics. The states of the model are susceptible, representing individuals who have not yet contracted the disease but are at risk; infected, representing individuals who are currently infected and can transmit the disease to susceptible individuals; and recovered, representing individuals who have recovered from the disease and are assumed to be immune and no longer susceptible. The differential equations specify the change in susceptible, infected, and recovered members of the population. You can now easily implement this model yourself (see exercises).

4.1.3 FITTING MODELS: STOCHASTICITY VERSUS NOISE

Grind includes an option to fit dynamical systems models. With `fit()`, based on the `modFit()` function from the FME package (Soetaert, Petzoldt, and Setzer 2010), one can estimate the model parameters given a dataset. These functions also provide confidence intervals and allow fixing parameters and bootstrap analysis. Fitting nonlinear dynamical systems models to data is an art in itself. For example, these methods can be very sensitive to the choice of initial values.

I will illustrate the use of `fit()` on three datasets created with the original Lotka–Volterra model from the previous section. The first dataset is the deterministic dataset, the data that follow directly from the code above. The second is created using a stochastic Lotka–Volterra model. I will explain how this works in the next section. The third is a deterministic dataset with measurement error. We will see that the last two cases are very different.

```
set.seed(1)
layout(matrix(1:4, 2, 2, byrow = TRUE))
p <- c(a = 1.1, b = .4, c = .1, d = 0.4)
# p is a named vector of parameters
s <- c(N = 10, P = 10) # s is the state
n <- 30
data_deterministic <- run(odes = LV, n, table = TRUE,
    timeplot = FALSE) # deterministic data
data_stochastic <- run(odes = LV, n, table = TRUE,
    after = "state<-state+rnorm(2,0,.1)",
    timeplot = FALSE) # add stochasticity
data_error <- run(odes = LV, n, table = TRUE, timeplot = FALSE)
data_error[,2:3] <- data_error[,2:3]+
    matrix(rnorm(2 * n, 0, 2), , 2) # measurement error
```

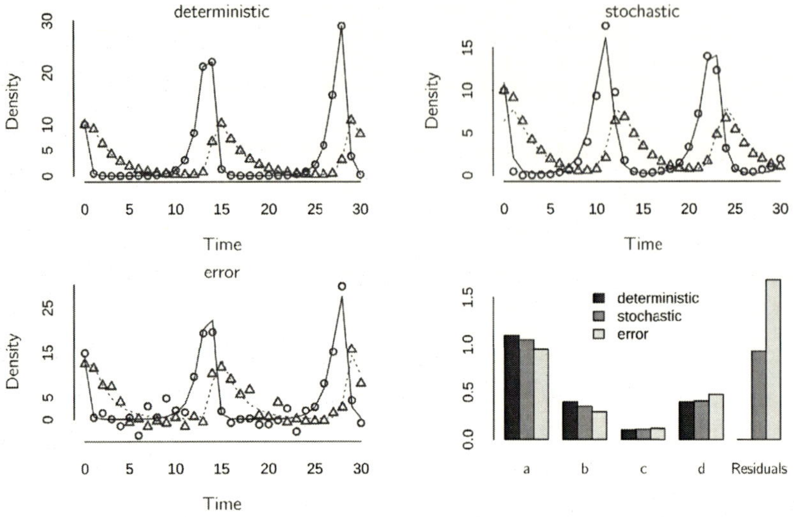

Figure 4.1. Fit of the Lotka–Volterra model on three types of data. The lines represent the fitted curves. In the stochastic case, noise is part of the system that affects the computation of the state at the next time step. In the error case, noise is a measurement error that does not affect the dynamics. Refer to the online version of the book for a color image.

```
#fit & plot:
s <- s * abs(rnorm(2, 1, 0.1)); s # start values
p <- p * abs(rnorm(4, 1, 0.1)); p # start values
f_deter <- fit(odes = LV, data_deterministic,
                main = 'deterministic')
f_stoch <- fit(odes = LV, data_stochastic,
                main = 'stochastic')
f_error <- fit(odes = LV, data_error, main = 'error')
pars <- matrix(c(f_deter$par[3:6], f_stoch$par[3:6],
    f_error$par[3:6]), ,3)
pars <- rbind(pars, c(summary(f_deter)$sigma,
    summary(f_stoch)$sigma, summary(f_error)$sigma))
barplot(t(pars), beside = TRUE, names = c('a','b','c','d',
    'Residuals'), args.legend = c(x = 13),
    legend.text = c('deterministic', 'stochastic', 'error'))
```

This results in figure 4.1.

Note that the error dataset looks very similar to the deterministic dataset because it contains only the measurement error (X is true score + error). The error does not affect the dynamics itself. In the stochastic case, the error (noise) is added to the states after each time step, which affects the dynamics. In figure 4.1, you can see that the positions of the waves change. In this well-chosen case, the fit is quite good in all three cases, and the parameter estimates are all quite close to the true values. Unfortunately, these results are quite unstable. You can do some testing yourself.

4.1.4 BACK TO THE CUSP

To illustrate how Grind can be used to perform bifurcation analysis, we go back to the cusp. Recall that the differential equation for the cusp is

$$\frac{dX}{dt} = -V'(X) = a + bX - X^3. \tag{4.4}$$

```
model <- function(t, state, parms){
  with(as.list(c(state, parms)),{
    dX <-  a + b * X - X^3 # cusp
    return(list(dX))
  })
}
p <- c(a = 0, b = 1); s <- c(X = .1); run(ymin = -1)
s[1] <- -.1; run(add = TRUE)
```

This code simulates two runs demonstrating bistability for $a = 0$ and $b = 1$.

A nicer way to demonstrate bistability in the time series is to make the system stochastic. This was done in chapter 3, section 3.5, by using a stochastic differential equation: $dX = -V'(X)dt + \sigma dW(t)$. Grind has a great trick for this. With the "after" parameter in the function call, we can add discrete events to the system. "After" can also be used to change parameter or state values after a certain amount of time or after some condition (see the manual). We use it here to add a random number sampled from a normal distribution, with a mean of 0 and a standard deviation of .4, to X. The best way to simulate this in Grind is using the Euler method with a small time step. The noise term should be corrected with $\sqrt{dt}$, as shown in the code. The Wikipedia page on stochastic differential equations will tell you more about the underlying ideas.

Stochastic resonance is a notable example of how noise, which is often considered undesirable or disruptive, can actually play a constructive role in certain systems, helping to reveal hidden patterns and structures that might otherwise remain obscured.

As can be seen in figure 4.2, the stochastic force causes spontaneous jumps between the two modes of the cusp. When noise or random fluctuations cause the entire equilibrium landscape of a dynamical system to become observable, it is often referred to as stochastic resonance. You

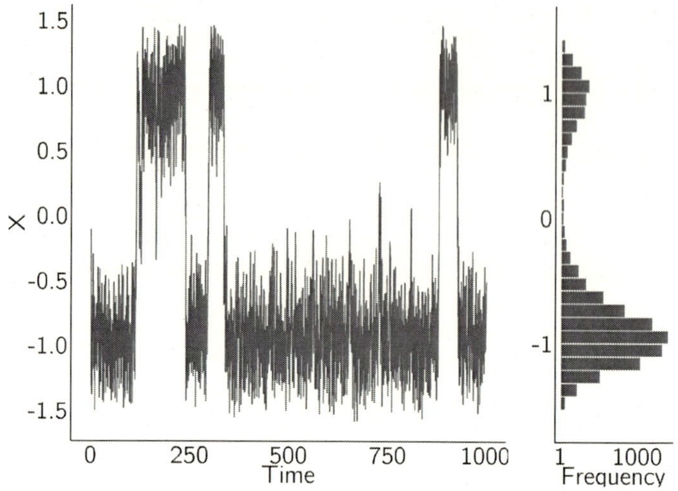

Figure 4.2. Spontaneous jumps in the cusp due to stochastics (noise). Due to stochastic perturbations, the system occasionally jumps over the maxima that separate the minima. Interestingly, when the noise is reduced, the time series tends to become trapped in a single equilibrium. Thus, increased noise helps reveal the overall equilibrium landscape. This phenomenon is known as stochastic resonance.

can see this by comparing the figure with one generated with a standard deviation of .1 or less.

```
layout(t(c(1,1,1,2)))
data <- run(table = TRUE, tmax = 1000, method = 'euler',
    step = .1, after = "state <- state +
    rnorm(1, mean = 0, sd = 0.4) * sqrt(tstep)",
    ymax = 2, ymin = -2, timeplot = FALSE)
plot(data, type='l', bty='n')
barplot(hist(data[,2], 30, plot = FALSE)$counts,
        xlab = "X", hor = TRUE)
```

4.1.5 BIFURCATION ANALYSIS

By combining the Grind functions `newton()` and `continue()` , we can perform bifurcation analysis. The `newton()` function finds stable and unstable fixed points, and the `continue()` function implements the parameter continuation of a steady state, providing a bifurcation diagram. It shows the change in equilibria when we change a parameter. This is what we did in chapter 2 for the logistic map, when we varied r and plotted the equilibria (see fig. 2.8).

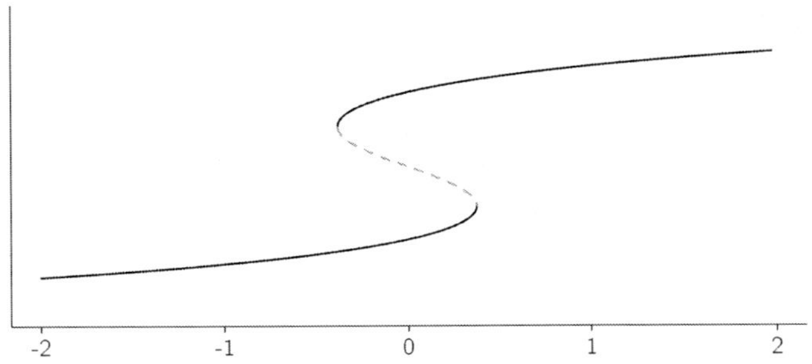

Figure 4.3. Hysteresis plot made with `newton()` and `continue()`. The function `newton()` finds an equilibrium, which is used in `continue()` to vary the normal variable a until a bifurcation point is found.

It is often necessary to run the combination of these two functions repeatedly, starting from different initial states. The code to create figure 4.3 is:

```
p <- c(a = 0, b = 1)
low <- newton(s = c(X = -1))
# finds a minimum starting from X = -1
continue(low, x = "a", y = "X", xmin = -2, xmax = 2, ymax = 2)
# Continue this steady state varying a
high <- newton(s = c(X = 1)) # again starting from X = 1
continue(high, x = "a", y = "X",
         xmin = -2, xmax = 2, ymax = 2, add = TRUE)
```

Another great tool in R is the deBif package by de Roos.[5] This is an R Shiny application that uses the same model specification and allows for a more interactive investigation. Given our previous model and the definition of s and p, we can run:

```
install.packages("deBif")
library(deBif)
phaseplane(model, s, p)
```

The `phaseplane()` function returns a time plot and the steady states. You can change parameters and initial states on the left side, and plot parameters on the upper right side (click on the two gears). The Steady States option is very useful as it shows the stable and unstable fixed

[5] https://cran.r-project.org/web/packages/deBif/deBif.pdf

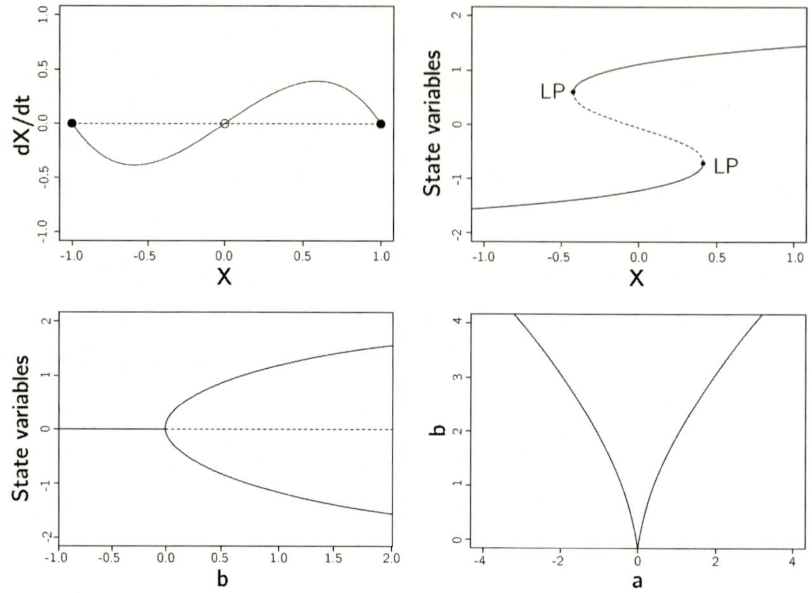

Figure 4.4. Output from Shiny app deBif. The last plot is a two-dimensional bifurcation diagram showing the bifurcation lines of the cusp in the a, b plane. This plot cannot be made with Grind.

points. Make sure that the minima and maxima of the plot axes are set correctly.

With

```
bifurcation(model, s, p)
```

you can create one- and two-parameter bifurcation diagrams (using the LP curve option, see figure 4.4). The two-parameter bifurcation diagram (bottom left) cannot be created in Grind. See the deBif help pages (with `??deBif`) for further instructions.

4.1.6 SPRUCE BUDWORM OUTBREAK MODEL

In chapter 3, section 3.4, I introduced the spruce budworm outbreak model. We will use this model later as a model of addiction. The bifurcation diagram can be made with:

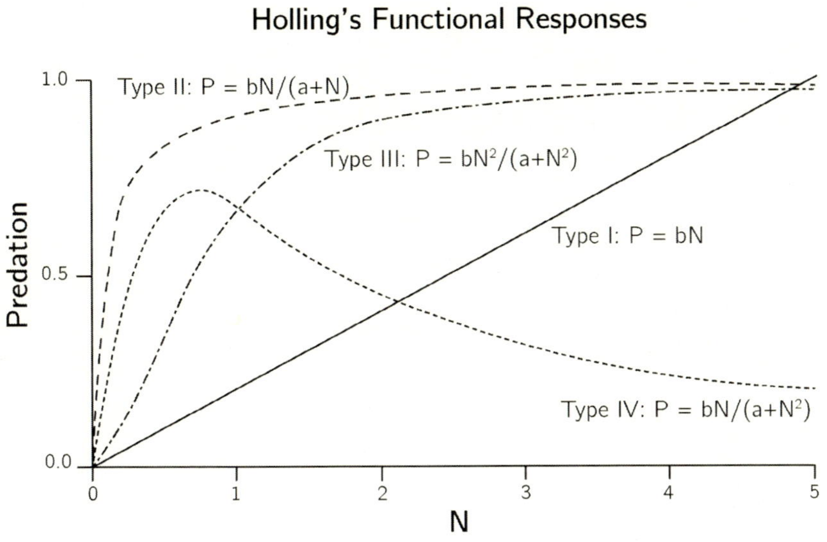

Figure 4.5. The Holling functional response models. Type III is used in the spruce budworm model.

```
spruce <- function(t, state, parms) {
  with(as.list(c(state, parms)), {
    du = r * u * (1 - u/q) - u^2 / (1+u^2)
    return(list(c(du)))
  })
}
state <- c(u = 0.5)
parms <- c(r = 0.4, q = 10)
bifurcation(spruce, state, parms)
```

Note that this predator–prey model consists of only one equation. There is no separate dynamic equation for the birds. The reason is that these budworm outbreaks happen in a few weeks. Birds do not reproduce on this time scale. The variables are reparameterized (see sec. 3.4). The predation term, in the original parameterization $-BN^2/(A^2+N^2)$, also has a logistic form that starts to accelerate at $N = A$ up to the maximum level B. The slow start A is used because birds only switch their diet to budworms when this population reaches a certain level (Ludwig, Jones, and Holling 1978). The fixed number of birds can only eat B budworms. This specific predation term is called the Holling type III model. All Holling types and their formulas are shown in figure 4.5.

4.1.7 EVALUATION OF ECOLOGICAL MODELING

Hundreds of extensions and variants of the Lotka–Volterra model have been proposed and studied.

Understanding the technical basics of dynamical systems theory is one thing, but actually building useful dynamical system models is quite another. Every term in each differential equation of a model needs some underpinning. These models make many assumptions, both implicit and explicit. The Lotka–Volterra model, for example, assumes that the prey population grows exponentially in the absence of the predator, that the predator population dies off with the prey population and does not switch to other prey species, that the response of the predator population to changes in the prey population is direct and not delayed, that there is no spatial component to the model, and that the rates of change of the populations are proportional to their sizes, to name just a few. These assumptions are widely debated in the biological literature (Abrams et al. 2000), and modifying these assumptions may have significant consequences.

For example, the original Lotka–Volterra model has no stable points, only limit cycles. While these cycles have been observed in nature, they are not overwhelmingly common. As we have seen, the dynamics of the system are significantly altered when prey growth is made dependent on prey density. This model has fixed-point equilibria instead of limit cycles.

Adding a spatial component can also make a big difference, as shown in the example of hypercycles in see section 5.2 (see Szostak, Wasik, and Blazewicz 2016 for a brief review). Adding more prey and predator species also makes a difference (Johnson, Mumma, and St-Laurent 2019). There are many interesting options for the predator term in the prey equation. Tyutyunov and Titova (2020) compare twelve trophic functions, alternatives to the Holling functional responses. The options are overwhelming. Biologists face a problem here that I discussed in chapter 1, section 1.4. Models easily become too complex. Recall that the traffic simulation models were extremely simple, yet sufficient to explain key phenomena.

An additional problem is that empirically testing all these different models is difficult. Although the quality of biological data is often superior to that of psychological data, biologists must also rely on the qualitative predictions of their models. Models in chemistry and especially physics can often be tested quantitatively. Transitions occur precisely at the predicted values of the control variables. Ecological models, much like those in psychology, do not allow for this level of prediction. This is a problem because if we can only test our model qualitatively (Are there limit cycles? What type of transitions can be detected? Is there hysteresis?), many model choices are not particularly relevant.

One of the most significant challenges in complex-systems research in the life sciences and psychology is constructing dynamical system models that effectively address these data-related issues.

A case in which this is less of an issue is the traffic example that I introduced in chapter 1. I asked you to play around with the online simulation. We now know the basics to better understand this model. The Wikipedia page on this model (Intelligent Driver Model) presents the equations, which are also coupled ordinary differential equations. The implementation in Grind of the simplest case looks like this:

```
model <- function(t, state, parms){
  with(as.list(c(state, parms)),{
    x <- state[1:n]
    v <- state[(n + 1):(2 * n)]
    dx <-  v # change in distance = speed
    delta_v <- v - m %*% v # difference in speed to next car
    s_alpha <- m %*% x - x - l # distance to next car
    s_alpha[n] <- 100 # front car has no car in front
    s_star <- s0 + v * T + v * delta_v / (2 * sqrt(a * b))
    # change in speed:
    dv <- a * (1 - (v/v0)^delta - (s_star/s_alpha)^2)
    return(list(c(dx, dv)))
  })
}
n <- 50
p <- c(l = 5, v0 = 30, T = 1.5, a = .73,
       b = 1.67, delta = 4, s0 = 2)
x_init <- (0:(n-1)) * (p['s0'] + p['l'])
v_init <- rep(0, n)
s <- c(x_init, v_init)
m <- diag(1, n, n); m <- rbind(m[-1,], 0) # order cars
# simulation with front car suddenly breaking at t = 150
data <- run(tmax = 300, timeplot = FALSE, table = TRUE,
            after = 'if (t==150) state[2*n] = 0')
matplot(data[,2:(n + 1)], type = 'l', bty = 'n',
        xlab = 'time', ylab = 'x')
```

The result of this simulation is shown in figure 4.6. Understanding the reasoning behind the differential equation is not so easy, but I want to make another point. The Wikipedia page gives parameters values with units (s, m/s, or m/s^2). One can also have dimension-free parameters (the acceleration exponent). This dimensional analysis is a crucial step in mod-

Dimensional analysis involves analyzing the dimensions of quantities to derive relationships and scaling laws, ensuring that the equations are consistent.

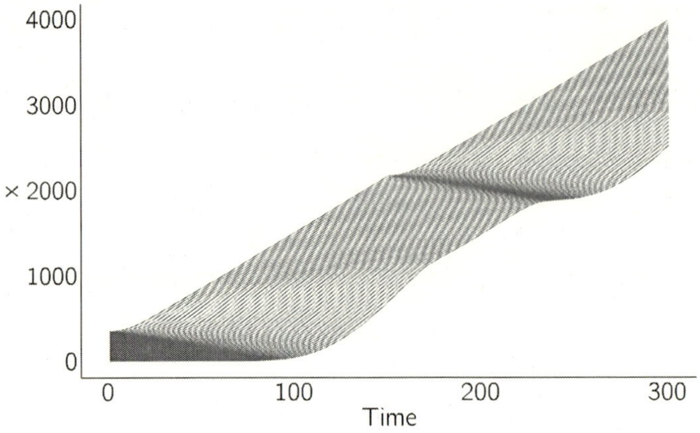

Figure 4.6. The traffic jam simulation. The top line represents the front car, which moves off immediately. Other cars are waiting for their turn. At $t = 150$, the first car suddenly breaks off, creating a traffic jam for the later cars. The effect of this disturbance is greater for the last car than for the first car. This simulated graph resembles the real data very well (see for example fig. 9 in Jusup et al. 2022).

eling in physics but a weak point in biological and especially psychological applications. This hampers the quantitative test of models.

4.3 Psychological Models

In this section, I present an overview of dynamical systems models in psychology, primarily in the form of systems of differential equations. Although the list is extensive, it is not exhaustive. It is important to explore different models and applications before embarking on your own modeling efforts.

4.2.1 RESPONSE TIME MODELS

Many dynamic models have been proposed in the study of speeded decision-making (Bogacz et al. 2006). The best-studied case is the two-alternative forced-choice task, where a stimulus is presented, and a choice must be made between two alternatives as quickly as possible. The stimulus could be an arrow pointing left or right. Most popular are accumulator models (fig. 4.7).

Accumulator models assume that noisy information is accumulated over time until a decision bound is reached and a motor response is initiated.

One way to model this process is with a single stochastic linear differential equation, called the drift-diffusion model (DDM), with I as the stimulus-driven input:

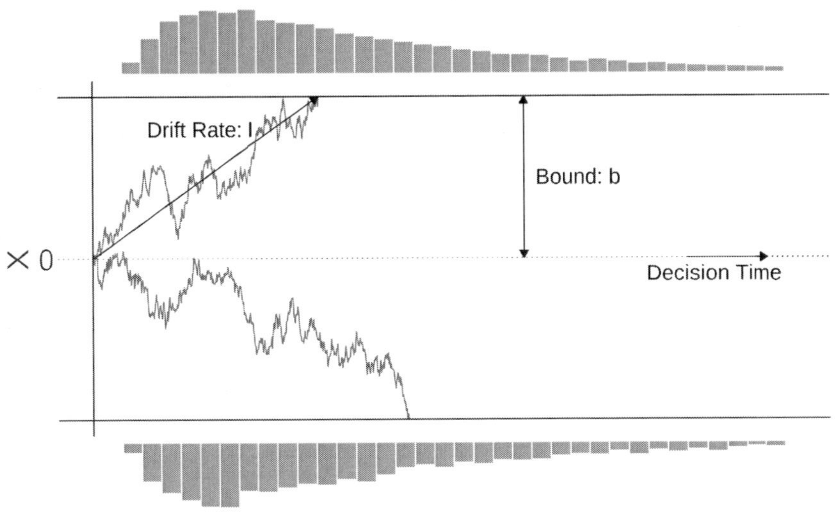

Figure 4.7. A stochastic accumulator model of speeded decision-making. Evidence accumulates in stochastic steps biased by the drift rate I (stimulus related). When one of the bounds is reached, a response is generated that may be incorrect if the bounds are too low.

$$dX = Idt + \sigma dW. \tag{4.5}$$

As before, dt is moved to the left-hand side of the equation. dW is white noise, normally distributed with 0 mean and with standard deviation σ (set to .1 by default). We start at 0, $X_{t=0} = 0$, assuming no bias for one of the choice alternatives.

The implementation of the model in Grind is quite simple. The trick is again in the run statement, which adds white noise after each step.[6] We stop the run when either the negative or positive bound is reached.

[6]Simulating this model correctly is more difficult than one might expect. I refer to Tuerlinckx et al. (2001) for a discussion of methods.

```
model <- function(t, state, parms) {
  with(as.list(c(state, parms)), {
    dX = I
    return(list(c(dX)))
  })
}
p <- c(I = .01); s <- c(X = 0)
bound <- 1
run(table = TRUE, method = 'euler', tstep = .1, tmax = 500,
    after = "state<-state+rnorm(1,mean=0,sd=0.1)*sqrt(tstep);
            if(abs(state)>bound) break", # stop at bound
    ymin = -bound, ymax = bound)
```

Accumulator models such as the drift-diffusion model explain the speed–accuracy trade-off. If we set our confidence bound higher, we are slower but more accurate.

The model explains observed response time and accuracy in terms of the underlying process parameters, drift rate, and confidence bound. By fitting the model to the data, we can determine whether slow responses are due to a low drift rate (low skill or difficult task) or a conservatively chosen bound.

A well-known extension of the drift-diffusion model is the Ornstein–Uhlenbeck model:

$$dX = (\lambda x + I)dt + \sigma dW. \tag{4.6}$$

For $\lambda < 0$ this process converges to I/λ (assuming $\sigma = 0$), while for $\lambda > 0$ it diverges. For the psychological interpretation, I refer to Bogacz et al. (2006). The simplest two-dimensional model is the race model:

$$\begin{aligned} dX_1 &= I_1 dt + \sigma W_1, \\ dX_2 &= I_2 dt + \sigma W_2. \end{aligned} \tag{4.7}$$

Now two independent processes run (race) to one positive bound. The first one to arrive wins. More biologically inspired models involve inhibition. The equations of mutual inhibition model are:

$$\begin{aligned} dX_1 &= (-k_1 X_1 - w_1 X_2 + I_1)dt + \sigma W_1, \\ dX_2 &= (-k_2 X_2 - w_2 X_1 + I_2)dt + \sigma W_2. \end{aligned} \tag{4.8}$$

Note that these are all linear dynamical systems that do not exhibit complex behavior. Examples of nonlinear alternatives are presented in Roxin and Ledberg (2008) and Verdonck and Tuerlinckx (2014) and discussed in Ratcliff et al. (2016).

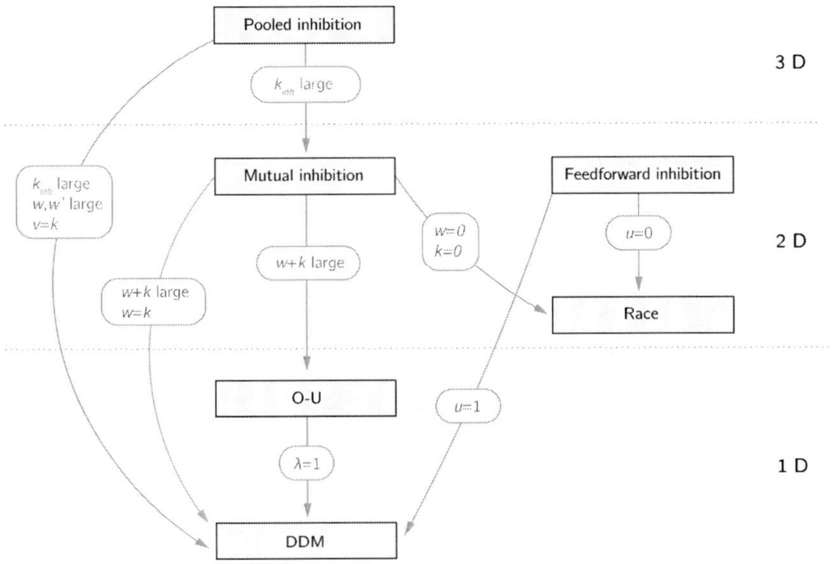

Figure 4.8. Relations between the main evidence accumulator models of decision-making. DDM is the drift-diffusion model which can be derived from the mutual inhibition model by setting $w = k$ and $w + k$ to a large value. *(Adapted from Bogacz et al. 2006).*

The relations between different accumulator models are summarized in figure 4.8. It shows that convenient models such as the drift-diffusion model can be derived by constraints on the parameters from more biologically realistic models, such as the pooled and mutual inhibition model.

4.2.2 DYADIC MODELS

The study of dyadic interaction lends itself to dynamic modeling. Dyadic interactions have been studied extensively in the field of caregiver–child interactions (Ainsworth et al. 2015). Here, we focus on a dyadic interaction in romantic relationships.

4.2.2.1 Romeo and Juliet

One type of model can be traced back to publications by Rapoport (1960) and Strogatz (1988). I follow the setup described by Sprott (2004). Note that it was intended as a toy model to demonstrate dynamical modeling.

The model is about the interactions between Romeo and Juliet, where R and J represent the feelings of Romeo and Juliet. The change in feelings is supposed to be a function of the feelings of both people:

$$\frac{dR}{dt} = aR + bJ,$$
$$\frac{dJ}{dt} = cR + dJ. \tag{4.9}$$

First note that the case of $b = c = 0$ resembles the exponential model with solutions $R = R_0e^{at}$ and $J = J_0e^{dt}$, which converge (to 0) or diverge (to infinity) depending on whether a and d are negative or positive. We will see a more sensible setup in the next model. Nevertheless, this system of coupled linear differential equations is surprisingly rich in behavior. With the signs of the parameters we define very different romantic styles. Strogatz (1988) distinguishes the following:

Divergence in this model, unbounded exponential growth of positive feelings, is an attractive concept but unrealistic, I'm afraid.

- Eager beaver: $a > 0$, $b > 0$, Romeo is encouraged by his own feelings as well as Juliet's.
- Narcissistic nerd: $a > 0$, $b < 0$, Romeo wants more of what he feels but retreats from Juliet's feelings.
- Cautious (secure) lover: $a < 0$, $b > 0$, Romeo retreats from his own feelings but is encouraged by Juliet's.
- Hermit: $a < 0$, $b < 0$, Romeo retreats from his own feelings as well as Juliet's.

Juliet may have her own style, which leads to complicated interactions. Sprott (2004) and other sources give an extended analytical treatment of this model. If you want to learn more about dynamical systems, you should study matrix algebra and its applications in linear dynamical systems. I have chosen to leave it out of this book because most psychological dynamical systems models are nonlinear. Here we just use Grind to test some cases. I give three examples with three different sets of parameter values. In the first case, the initial mutual interest fades; in the second case, the relationship fizzles out after some ups and downs; and in the third case, the couple ends up in a cycle of hate and love.

Systems of linear differential equations can be solved analytically, and the behavior of the equilibria can be characterized by the eigenvalues. Some knowledge of matrix algebra is required to understand this.

```
model <- function(t, state, parms) {
  with(as.list(c(state, parms)), {
    dR <- a * R + b * J
    dJ <- c * R + d * J
    return(list(c(dR, dJ)))
  })
}
layout(matrix(1:6, 3, 2, byrow = TRUE))
p <- c(a = -1, b = 1, c = .5, d = -1) # parameters
s <- c(R = 0.1, J = .1)
run()
plane(portrait = TRUE, ymin = -1, xmin = -1, grid = 3,
      vector = TRUE, legend = FALSE)
p <- c(a = -.2, b = -1, c = 1, d = 0) # parameters
run(ymin = -.2, legend = FALSE)
plane(portrait = TRUE, ymin = -1, xmin = -1, grid = 2,
      tstep = .001, legend = FALSE)
p <- c(a = -.1, b = -1, c = 1, d = 0.1) # parameters
run(ymin = -.2, legend = FALSE)
plane(portrait = TRUE, ymin = -1, xmin = -1, grid = 3,
      tstep = .001, legend = FALSE)
```

The lines in the phase plots in figure 4.9 are the nullclines. In linear dynamical system, nullclines are straight lines. Where they intersect, stable or unstable fixed points can occur. Depending on the angle between the nullclines, we get a fixed point (first two cases), a limit cycle (last case), or divergence (not shown).

Nullclines are the curves for which the time derivatives of the behavioral variables R and J are 0.

Rinaldi (1998) proposed an extension and a constraint to the model that makes it a bit more realistic and easier to study. The basic equation is now $dR/dt = -aR + bJ + A_J$, where a is interpreted as a forgetting parameter (constrained to be positive) and A_J is the attractiveness of the Juliet. In this case, a necessary and sufficient condition for asymptotic stability (i.e., having a fixed point) is that $ad > bc$.

Rinaldi also considers the case of a population of heterosexual men and women with different levels of attractiveness. The idea is that a man and a woman will leave their current partners and bond together when both reach a more optimal level of love. Rinaldi analyzes the conditions under which the population reaches a stable state. This marriage assignment problem, as it is called, is an example of a famous problem in optimization theory known as an assignment problem. The goal is to find a stable assignment of men to women, such that no man and woman prefer each other to their

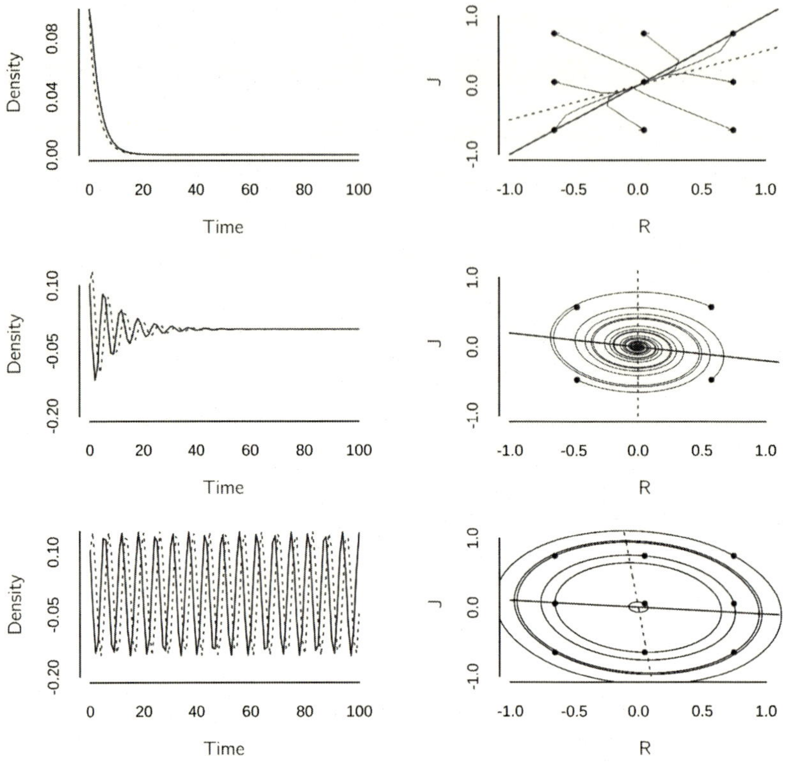

Figure 4.9. Three different love affairs between Romeo and Juliet. Refer to the online version of the book for a color image.

current partners (Gale and Shapley 1962). We use such algorithms to assign students to master tracks in our educational program.

Some other advanced variations of this model have been proposed. In these papers the analysis of the mathematical properties of the model gets much more attention than the psychological theory. It is often unclear what, exactly, the variables are and what the reasoning behind certain model assumptions. The work of Murray and Gottman, discussed in the next section, is more interesting in this regard.

4.2.2.2 The Mathematics of Marriage

The model of marriage developed by the psychologist John Gottman and the mathematical biologist James Murray (2002) is firmly grounded in psychological theory and data. The main phenomenon that inspired this modeling work is Gottman and Levenson's (1992) finding that the patterns of interaction between couples, when discussing a major area of ongoing disagreement in their marriage, are predictive of divorce.

The model consists of two coupled difference equations, but I present it in the form of differential equations.[7]

$$\begin{aligned} \frac{dW}{dt} &= I_w(H, a, b) - r_w W + W_e, \\ \frac{dH}{dt} &= I_h(W, a, b) - r_h H + H_e, \end{aligned} \tag{4.10}$$

where the influence functions I_w and I_h are defined as:

$$I(x, a, b) = \frac{sgn(x)}{1 + e^{a(|x|-b)}}. \tag{4.11}$$

I made up this flexible function to allow for very different forms of influence (as we will see below). When both influences are 0 ($a = -8, b = -\infty$), the state or mood of the wife (W) and the husband (H) converge to W_e and H_e, with rates r_w and r_h, respectively. W_e and H_e are the uninfluenced steady states of mood when the spouses do not interact.[8]

However, if the influence function ($a = -8, b = 0$) is such that a positive mood in one spouse provokes a positive mood in the other, while a negative mood provokes a negative mood, we expect a negative and a positive equilibrium depending on the initial states and uninfluenced steady-state values.

Another more complex influence function ($a = -8, b = 1$) assumes that only extreme mood states influence the other spouse.

This is implemented with:

[7] Difference equations were used in the original model because the data consist of turn takings in a conversation. This, however, does not lead to qualitative different results. With `method = 'euler'` and a change in r_w and r_h the difference model can be constructed.

[8] I follow the definition and notation of the original source, but this model is clearly not restricted to heterosexual relationships.

```
model <- function(t, state, parms) {
  with(as.list(c(state, parms)), {
    dW <- influence(H,a,b) - rw * W + We
    dH <- influence(W,a,b) - rh * H + He
    return(list(c(dW, dH)))
  })
}
layout(matrix(1:9, 3, 3, byrow = TRUE))
par(mar=c(4,4,1,2))
influence <- function(x, a = -8, b = 1)
  sign(x) / (1 + exp(a * (abs(x) - b)))
p <- c(rw = .6, rh = .6, We = .18, He = -.18, a = -8, b = Inf)
s <- c(W = 0, H = 0)
for(b in c(Inf, 0, 1)){
    p['b'] <- b
    curve(influence(x, -8, b), -3, 3,
          xlab = 'W', ylab = 'H', lwd = 2)
    plane(xmin = -2.5, xmax = 2.5,
          ymin = -2, ymax = 2, legend = FALSE)
    for(i in seq(-2, 2, by = .25))
        newton(s = c(W = i, H = i), plot = TRUE)
    for (i in 1:100)
        run(state = c(W = rnorm(1,0,.5), H = rnorm(1,0,1)),
            tmax = 50, ymin = -2, ymax = 2, add = (i>1),
            legend = FALSE)
}
```

This results in figure 4.10.

The three influence functions are shown in the first column, the nullclines in the second, and a series of runs from random initial states in the third. The first case shows that the moods converge to W_e and H_e, when there is no mutual influence. The second has two equilibria, with an unstable fixed point in the middle, while the last case has five fixed points, three of which are stable.

I have kept the influence functions and most parameters the same for both spouses, but this is not necessarily the case. It is possible to derive the equilibria analytically and to determine the stability of these equilibria (Gottman et al. 2002). They also propose a two-stage procedure for fitting the model to data consisting of positive and negative speaker interactions using a bilinear influence function. A more advanced statistical approach is proposed in Hamaker, Zhang, and van der Maas (2009). For a related model for the interaction between therapist and client, see Tschacher and Haken (2019).

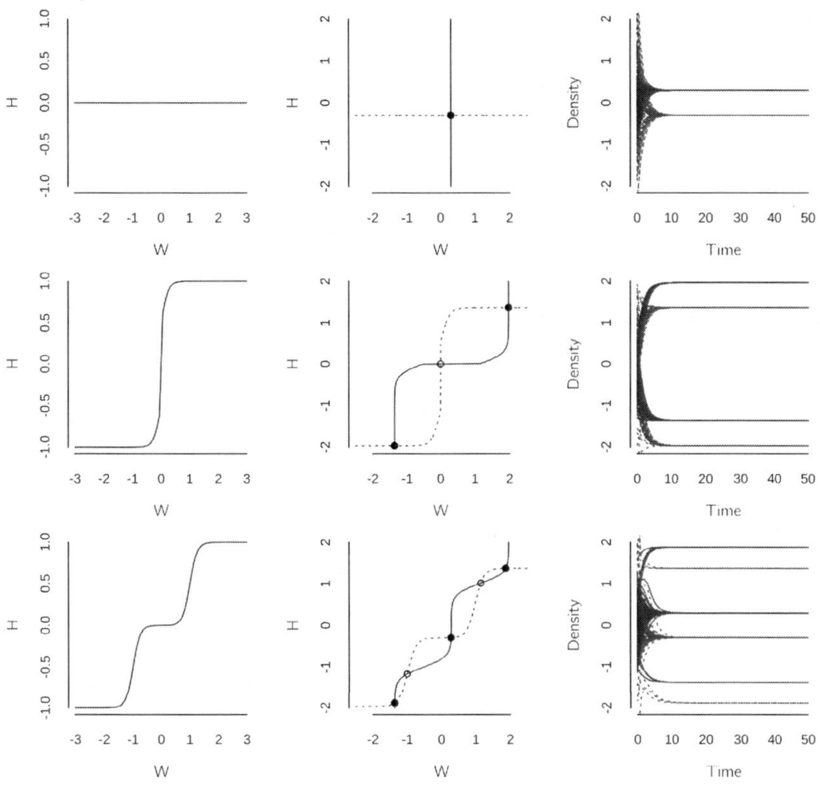

Figure 4.10. Qualitative difference marriage equilibrium landscapes depending on the form of the influence function. In the first (top), they simply have no influence and both partners converge to their uninfluenced steady states of mood. In the second (middle), the response to the partner's mood is extreme, resulting in either a positive or negative mutual state. In the last case (bottom), the response to low positive or negative moods is close to 0 but extreme at higher levels. Now there are three stable states. Refer to the online version of the book for a color image.

4.2.3 THE VAN GEERT MODELS

In a series of papers, Paul van Geert proposed dynamical systems models for developmental processes (Den Hartigh et al. 2016; van Geert 1998, 1991).

It is thought that cognitive and language abilities grow over time in an autocatalytic process constrained by a limited capacity, similar to the logistic growth of populations.

Van Geert has proposed many different models, but I will give just one example. Van Geert (1991) introduced a system of two coupled difference equations to model where the growth rate of one cognitive ability depends on the level of another cognitive ability:

$$X_{t+1} = (a - bY_t)X_t - \frac{aX_t^2}{K},$$
$$Y_{t+1} = (c - dX_t)Y_t - \frac{cY_t^2}{K}. \tag{4.12}$$

This can be implemented as follows:

```
model <- function(t, state, parms) {
  with(as.list(c(state, parms)), {
    dX <- (a + b * Y) * X - a * X^2 / K
    dY <- (c + d * X) * Y - c * Y^2 / K
    return(list(c(dX, dY)))
  }) }

layout(matrix(1:4,2,2))
# Set parameter values and run the model:
p <- c(K = 1, a = 0.4, b = -0.05, c = .4, d = -0.15)
s <- c(X = 0.01, Y = 0.01)
run(method = "euler", tstep = 1)
plane(portrait = TRUE, grid = 4)
p <- c(K = 1, a = 0.05, b = -0.1, c = 0.05, d = -0.09)
s <- c(X = 0.0126, Y = 0.01)
run(tmax = 1500, method = "euler", tstep = 1)
plane(portrait = TRUE, grid = 4)
```

So, there are basically two outcomes: either both grow, or one grows and suppresses the other (see fig. 4.11). Note that the method is set to "Euler" to simulate difference equations.

4.2.4 THE PÓLYA URN MODEL OF THE THIRD SOURCE

Another type of discrete dynamical systems model is the Pólya urn model, which is relevant to understanding nonlinear developmental processes in psychology. Molenaar, Boomsma, and Dolan (1993) propose the third-source hypothesis. Based on a series of studies, Gärtner (1990) concluded that 70–80% of the variation in body weight in inbred mice appears to be due to a third component that generates biological variability in addition to genetic and environmental influences.

The third-source hypothesis proposes that the development of complex living systems is influenced by three sources of variation: genetic variation, environmental variation, and self-organizing processes.

A simple, and in my opinion insightful, dynamical model for this effect is the Pólya urn model (Mahmoud 2008). In this discrete dynamical model, we add marbles to an urn containing some red and blue marbles. We could start with two blue and one red marbles. We randomly take out a marble.

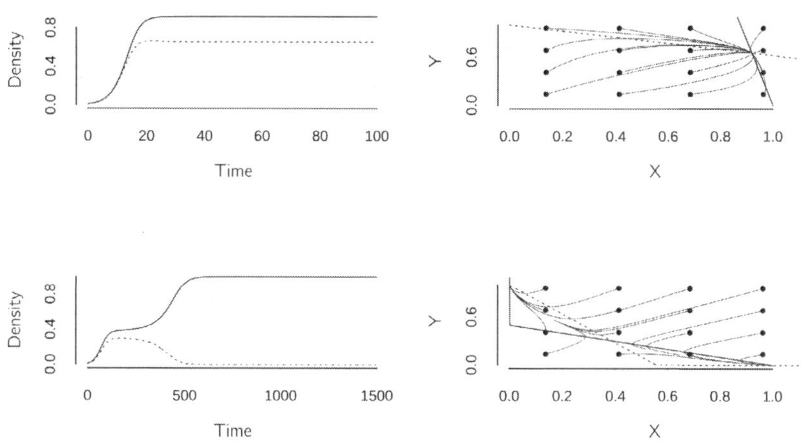

Figure 4.11. The development of two interrelated cognitive abilities in one of the van Geert models. In the left case they coexist; in the right case one ability suppresses the other.

If it is blue, we put it back with another blue marble. If it is red, we put it back and add a red marble. Initially, $p(blue) = 2/3$, but what will happen to that probability over time when we have more and more marbles? Think about it for a moment.

My intuition was simply wrong, and in my experience, this is true for the vast majority of people. What happens is shown in figure 4.12. Each time you run the process, $p(blue)$ reaches a stable state, but the value of that state is random. What happens is that early (random) samples have a huge influence on the long-term dynamics. This creates a Matthew effect.

The Matthew effect says that the rich get richer and the poor get poorer.

Savi et al. (2019) provide a developmental interpretation. Imagine a child receives a tennis racket for her birthday. First, she practices the backhand twice at home, but incorrectly. Then, during her first tennis lesson, her trainer demonstrates the correct backhand. She now has three experiences, two incorrect and one correct. Now, suppose her backhand development is based on a very simple learning schema. Whenever a backhand return is required, she samples from her earlier experiences, and the sampled backhand is then added to the set of earlier experiences. Thus the accumulation of experiences follows the Pólya urn scheme. While she has the potential to become a tennis master, her twin sister, who had less fortunate initial experiences, decides to quit tennis lessons within the first year. This model

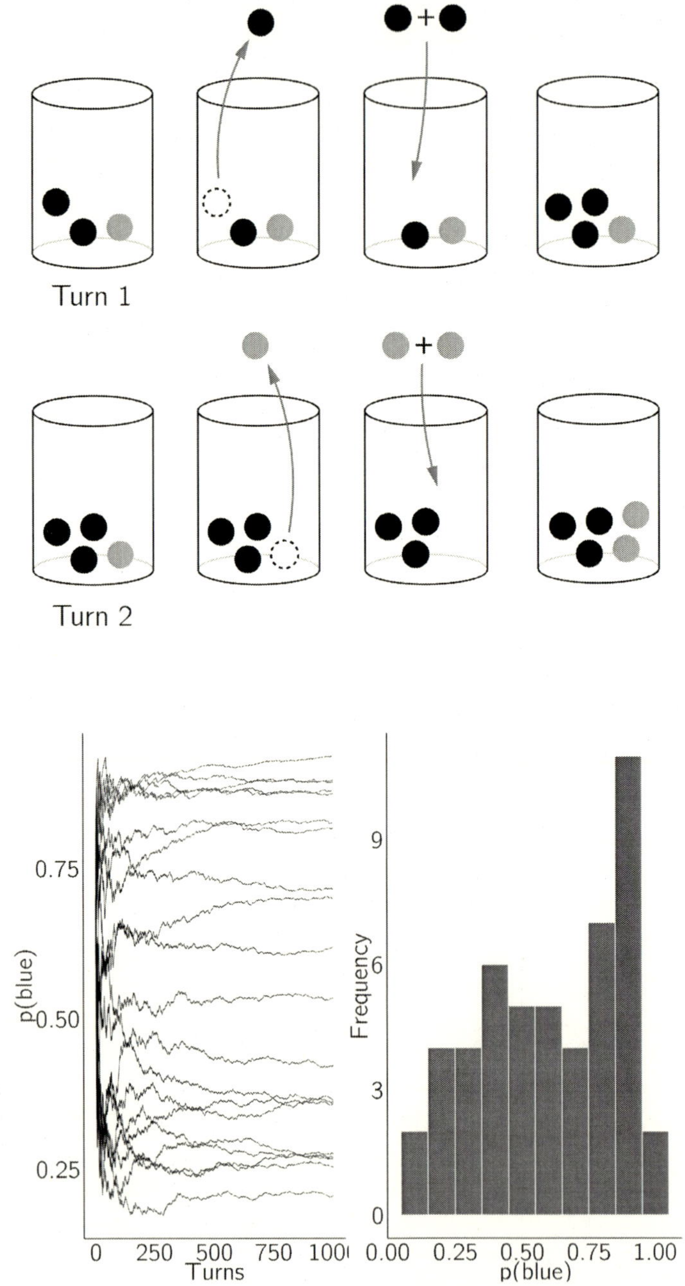

Figure 4.12. The Pólya urn model. A random marble is sampled and placed back with an extra marble of the same color. The evolution of the probability of picking a blue marble is unpredictable and converges to a random number. This mechanism may play a role in the Matthew effect. Refer to the online version of the book for a color image.

is consistent with many developmental theories (e.g., the critical period hypothesis), but these theories lack a formal approach.

4.2.5 THE PANIC MODEL

In recent years, we have been working on a model of panic disorder (Robinaugh et al. 2019). In theories of panic disorder, there is a reinforcing feedback loop between arousal and perceived threat. When an increase in arousal is perceived as a threat (e.g., a heart attack), arousal increases further. This "vicious cycle" results in a panic attack (D. M. Clark 1986). Thus, these theories posit two causal effects: an effect of perceived threat on arousal and an effect of arousal on perceived threat.

We will further assume that the effect of perceived threat on changes in arousal is essentially linear while the causal effect of arousal on perceived threat is nonlinear (S-shaped). For the argument, see Robinaugh et al. (2019). It could be argued that both are nonlinear, but this does not fundamentally change the qualitative behavior of the model. The central part of the model consists of two coupled differential equations:

$$\begin{aligned} \frac{dA}{dt} &= -A + bT, \\ \frac{dT}{dt} &= \frac{1}{1 + e^{-\alpha(A+\beta)}} - T. \end{aligned} \tag{4.13}$$

This looks a bit like the Romeo and Juliet model, but now the effect of arousal A on the change in perceived threat T is a logistic function that starts at 0 and grows to 1. The location is determined by β, and the acceleration or steepness is determined by α.

An implementation and simple illustration is:

```
model <- function(t, state, parms) {
  with(as.list(c(state, parms)), {
    dA <- -A + b * T
    dT <- -T + 1/(1 + exp(-alpha * (A + beta)))
    return(list(c(dA, dT)))
  })
}
```

```
p <- c(b = 1, alpha = 12, beta = -.7)
s <- c(A = 0, T = 0)
# arousal increase for time t in 20:30, leads to panic,
# which after some time ('30 min') disappears
layout(1:2)
plane(vector=TRUE, xmin = 0, ymin = 0,
      xmax = 1, ymax = 1.1, legend = FALSE)
newton(s  c(A = 0, T = 0), plot = TRUE)
newton(s  c(A = 0.8,T = .8), plot = TRUE)
newton(s  c(A = 1,T = 1), plot = TRUE);
run(after="if(t>20&t<30)state[1]<-1;
  state<-state+rnorm(2,mean=0,sd=0.1)")
```

The β parameter is set so that the nonpanic mode dominates, but the panic mode is present (a metastable state). This state can be easily disturbed (see plane). For $20 < t < 30$, arousal is set to a high value, resulting in a high perceived threat. But because we also added some noise to both processes, after some time both arousal and perceived threat jump back to low values.

This dynamic of this model is the cusp, as can be checked with (see fig. 4.15):

```
p <- c(b = 1, alpha = 12, beta = -.5)
start <- newton(s = c(A = .1,  T= .1))
continue(start, x = "beta", y = "T",
         xmin = -1, xmax = 1, ymax = 1)
continue(start, x = "alpha", y = "T",
         xmin = -1, xmax = 20, ymax = 1)
start <- newton(s = c(A = 1, T = 1))
# finds a minimum starting from X = -1
continue(start, x = "alpha", y = "T",
         xmin = -1, xmax = 20, ymax = 1, add = TRUE)
```

In Robinaugh et al. (2019), this model is extended with other processes, such as arousal and escape schemes, that operate on the parameters of the basic model. These are slower processes that are modeled on different time scales.

The difference in time scales refers to the different rates at which system components or processes evolve, affecting the overall behavior of the system.

4.2.6 NEURAL MODELS: VAN DER POL AND DIFFERENT TIME SCALES

In the panic model, we touched on differences in time scales. Time scales are critical to understanding and managing complex systems because they

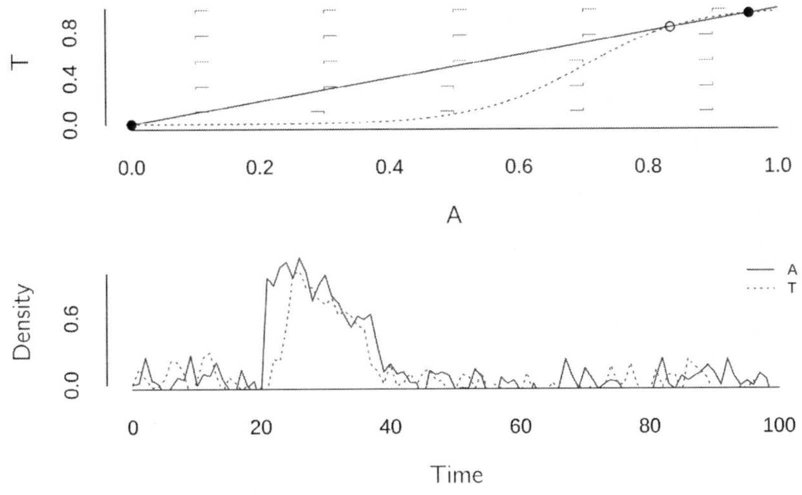

Figure 4.13. The panic model. Arousal is set high between time is 20 and 30, but panic persists due to the hysteresis effect. Eventually, due to noise, it escapes from the metastable attractor at $A = T = 1$.

allow fast dynamics to be separated from slow dynamics, thereby simplifying analysis and modeling. I will explain this further in the context of the van der Pol oscillator, which has many interesting applications.

Imagine taking the cusp equation $dX/dt = a + bX - X^3$, with $b = 1$, such that we have hysteresis. But now we make a, or actually da/dt, a function of X: $da/dt = -\varepsilon X$, where ε is small constant. If we set ε to .05, a changes twenty times more slowly than X. What happens now is that with $X = 1$ and $a = 0$, a decreases up to the point where X jumps to a negative value. Now a increases, resulting to a new jump to a positive value of X. And this loop will continue endlessly.

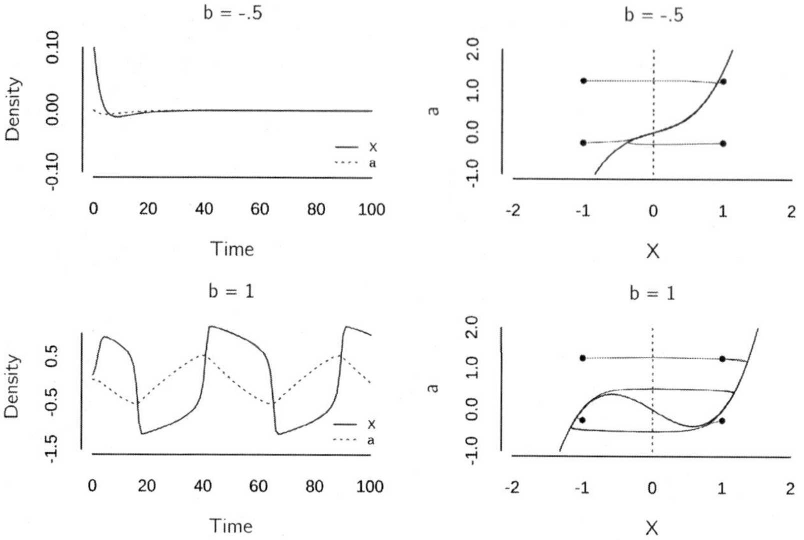

Figure 4.14. Two runs of the van der Pol oscillator. For high b, this system oscillates between the two stables states of the cusp. The black dots represent different initial states.

```
model <- function(t, state, parms){
  with(as.list(c(state, parms)),{
    dX <-  a + b * X - X^3 # cusp
    da <- -e * X
    return(list(c(dX, da)))
  })
}
s <- c(X = .1, a = 0) # initial state and parameter values
layout(matrix(1:4, 2, 2, byrow = TRUE))
p <- c(e = .05, b = -.5)
run(ymin = -.1, main = 'b = -.5', legend = FALSE)
plane(xmax = 2, ymin = -1, ymax = 2, xmin = -2,
      portrait = TRUE, grid = 2, main= 'b = -.5')
p <- c(e = .05, b = 1)
run(ymin = -1.5, main = 'b = 1', legend = FALSE)
plane(xmax = 2, ymin = -1, ymax = 2, xmin = -2,
      portrait = TRUE, grid = 2, main= 'b = 1')
```

The plots (fig. 4.15) illustrate this behavior. For $b < 0$, X converges to a fixed point. For $b > 0$, we see cyclic jumps up and down. This oscillator is basically the famous van der Pol oscillator, originally written in the form:

$$\frac{d^2X}{dt^2} = \mu\left(1 - X^2\right)\frac{dX}{dt} + x. \tag{4.14}$$

Such a second-order differential equation can be rewritten in the form of a first-order system of multiple equations, which is the form required for Grind.[9] This model form is of (neuro)psychological interest because it relates to the FitzHugh–Nagumo model for neuronal excitability (Izhikevich and FitzHugh 2006)[10]:

$$\begin{aligned}\frac{dV}{dt} &= V - \frac{V^3}{3} - W + I,\\ \frac{dW}{dt} &= .08(V + .7 - .8W).\end{aligned} \tag{4.15}$$

The equation for V, the membrane potential, has a cubic nonlinearity that allows regenerative self-excitation via positive feedback. W, a recovery variable, provides linear negative feedback. I represents the input. The main phenomena in this model are shown in figure 4.15.

This model is for a single neuron. Crucial is that second equation is a slow process. Time-scale effects also play an important role in learning in neural networks. In most neural networks, there is a fast equation for updating neuron activities and a much slower equation for updating the connection strengths.

Other applications of the van der Pol model concern extensions of the Haken–Kelso–Bunz (HKB) model (see sec. 5.4), multistable perception (Fürstenau 2014), developmental processes (Molenaar and Oppenheimer 1985), and bipolar disorder (Daugherty et al. 2009). One case where it seems especially useful is in modeling the wake–sleep cycle (Forger, Jewett, and Kronauer 1999).

4.2.7 ANALOGICAL MODELING: BUDWORMS AND BEERS

If we create a dynamic model from the ground up, there's a significant chance that we might not completely grasp its intricacies. We have seen that some very simple models already show amazingly complex behavior.

One approach to cope with these issues is analogical modeling, or basically copying models. For instance, we used the Ising model to model attitudes and the mutualistic Lotka–Volterra model to model intelli-

[9] The rewriting is based on the Liénard transformation.

[10] The FitzHugh–Nagumo model is itself a simplified version of the famous Hodgkin–Huxley model, which consists of four differential equations and models the activation and deactivation dynamics of a spiking neuron in more detail.

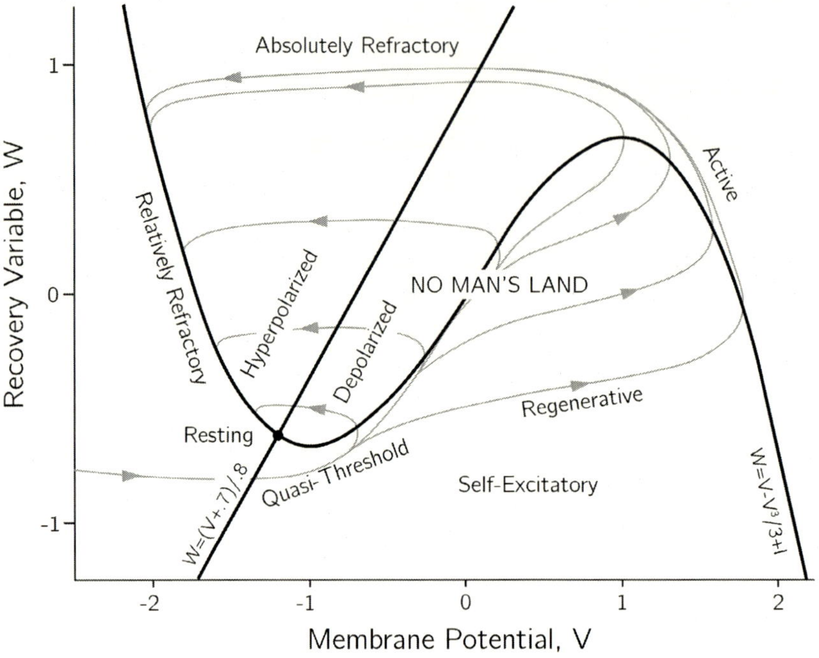

Figure 4.15. The dynamics of the FitzHugh–Nagumo model. The horizontal axis represents the membrane potential (V), which is the voltage across the membrane of a neuron. The vertical axis represents the recovery variable (W). The V-nullcline (where $dV/dt = 0$) and the W-nullcline (where $dW/dt = 0$) cross at the unstable resting point. The lines with arrows represent typical trajectories. In the depolarized state, the membrane potential is higher than at rest, potentially leading to an action potential. In the hyperpolarized state, the membrane potential is lower than at rest. In the self-excitatory and regenerative phases, the system can increase its own activity without external input. The active region refers to a state in which the neuron is actively firing. Absolute and relative refractory are periods following an action potential when a neuron cannot fire again.

gence. Both are explained in chapter 6. Here, I will use addiction as an example, focusing on a selection of key phenomena (and for now ignoring many others).

We have reviewed existing formal models of addiction in van den Ende et al. (2022). Most of these models are quite complicated. I want the model of the individual addict to be as simple as possible. The reasons for this will become clear in chapter 7 when we include social effects (Boot et al., submitted for publication). The key phenomena are that initiation, cessation, and relapse are often discontinuous processes. The verbal theories we adapt are dual-process models in which an automatic process of using more and more is controlled by a non-automatic process, self-control.

Instead of creating our own model, we look for well-studied models in other sciences, which led me to the spruce budworm model:

$$\frac{dN}{dt} = r_b N \left(1 - \frac{N}{K}\right) - \frac{BN^2}{A^2 + N^2}. \qquad (4.16)$$

But now we interpret the variables and parameters as follows. We first assume N is the number of drinks you consume. The time scale is a day or an evening (depending on when you have your first beer). K is the upper limit of drinks you can take, either because of lack of availability or, worse, because you just collapse. r_b is the addiction sensitivity. If this is too low ($r_b < 0$), the 0 state is stable. The logistic function seems to be a reasonable choice. Drinking might start off slow, then accelerate, and level off at K. This happens when there is only an autonomous process. The second term, the predator term, is now interpreted as self-control. This is not something that changes on the time scale of a day, so, as in the case of the birds, a second equation is not required. A (or actually $1/A$) is a responsiveness parameter, the number of drinks at which self-control is activated, which may not be at the first or second beer. B is the maximum level of self-control. As in the original model, this term is a Holling type III form (see fig. 4.5). We could also insert a Holling type IV form, with the idea that self-control deteriorates after too many drinks. Depending on the values of the parameters, one may not drink at all ($r_b < 0$), drink at a recreational level, or have an "outbreak" to heavy use.

The advantage of this type of analogical modeling is that we already know everything about the model. We know it is a cusp, and we have already made the bifurcation diagram. There are also disadvantages or ambiguities.

First, the definition of N is imprecise. Is it the blood alcohol concentration, the number of drinks, or some other quantity?

Second, the choice of a logistic function for the autonomous part seems reasonable but is not derived from first principles. One could also assume a linear function with a ceiling at K.

Third, and relatedly, the self-control function is also not derived from first principles. An additional problem is that we cannot measure this term directly (Duckworth and Gross 2014).

Fourth, this model may not work for all addictions or should be adapted to specific cases. An example is smoking. For smoking, the intermediate recreational state is very unstable (Epskamp et al. 2022), and the autocatalytic effect described by the logistic equation seems less appropriate. For alcohol, the Holling type IV seems to be a good choice for the self-control term as alcohol directly impairs brain regions involved in self-control (Remmerswaal et al. 2019). For gambling, Holling Type III may be sufficient.

Fifth, processes at other time scales are missing. The model seems to work well for the time scale of a day or an evening. Other relevant time scales are minutes (direct effect of alcohol intake on the brain), weeks (abuse is often concentrated on weekends), and months. On time scales of months or even years, the parameters r_b, K, and B may change. For example, experienced drinkers can drink more. Also, the r_b, addiction sensitivity, may slowly increase over time. This can be taken into account with additional equations. Furthermore, K, A, and B could change as an effect of the social network. Nondrinkers might increase A, while other users in the social network might increase K (availability). These modeling issues are serious but also very interesting (Dongen et al. 2024).

Ambiguities in our thinking about psychological systems come to light in the process of building concrete mathematical models.

4.2.8 CASCADING TRANSITIONS IN MULTIFIGURE MULTISTABLE PERCEPTION

4.2.8.1 Interacting Cusps

In section 4.3, we studied the van der Pol oscillator. In that model the normal variable of the cusp was itself a dynamic variable $\frac{da}{dt} = -\varepsilon X$. Instead of a linear equation, we could also use a cusp. We then get:

$$\begin{aligned} \frac{dX}{dt} &= aY + bX - X^3, \\ \frac{dY}{dt} &= cX + dY - Y^3. \end{aligned} \tag{4.17}$$

This model, first proposed by Kadyrov, was analyzed in detail by Abraham et al. (1991). We can study this model in Grind by specifying:

```
model <- function(t, state, parms) {
  with(as.list(c(state, parms)), {
    dX <- a * Y + b * X - X^3
    dY <- c * X + d * Y - Y^3
    return(list(c(dX, dY)))
  })
}
```

Depending on the choice of the parameters and initial values, many different things can happen. Abraham et al. (1991) created bifurcation diagrams to summarize the qualitatively different regimes. We will restrict ourselves to the case where $b = d = 1$, and a and c are varied. The bifurcation diagrams and associated phase planes are shown in figure 4.16.

The phase planes of figure 4.16 can be made with:

```
layout(matrix(1:4, 2, 2))
s <- c(X = 0, Y = 0)
for(i in c('a','b','c','d'))
{
if (i == 'a') p <- c(a = .3, b = 1, c = .3, d = 1)
if (i == 'b') p <- c(a = .6, b = 1, c = .6, d = 1)
if (i == 'c') p <- c(a = 1, b = 1, c = 1, d = 1)
if (i == 'd') p <- c(a = 1, b = 1, c = -1, d = 1)
plane(tstep = 0.5, portrait = (i == 'd'), xmin = -2, ymin = -2,
      xmax = 2, ymax = 2, legend = FALSE, grid = 2,
      main = paste("Case ", i))
      # make a phase portrait (Fig 1c)
if (i != 'd') for(i in 1:200)
    newton(c(X = runif(1,-2,2), Y= runif(1,-2,2)), plot = TRUE)
else newton(c(X = 0, Y = 0), plot = TRUE)
}
s <- c(X = 0.1, Y = .1)
p <- c(a = 1, b = 1, c = -1, d = 1)
run(tmax = 20, tstep = 0.1, ymin = -2, ymax = 2)
# Kadyrov oscillator
```

The last three lines of this code show the Kadyrov oscillator (fig. 4.17).

If we simplify this analysis a bit to stable fixed points only, we see three regimes:

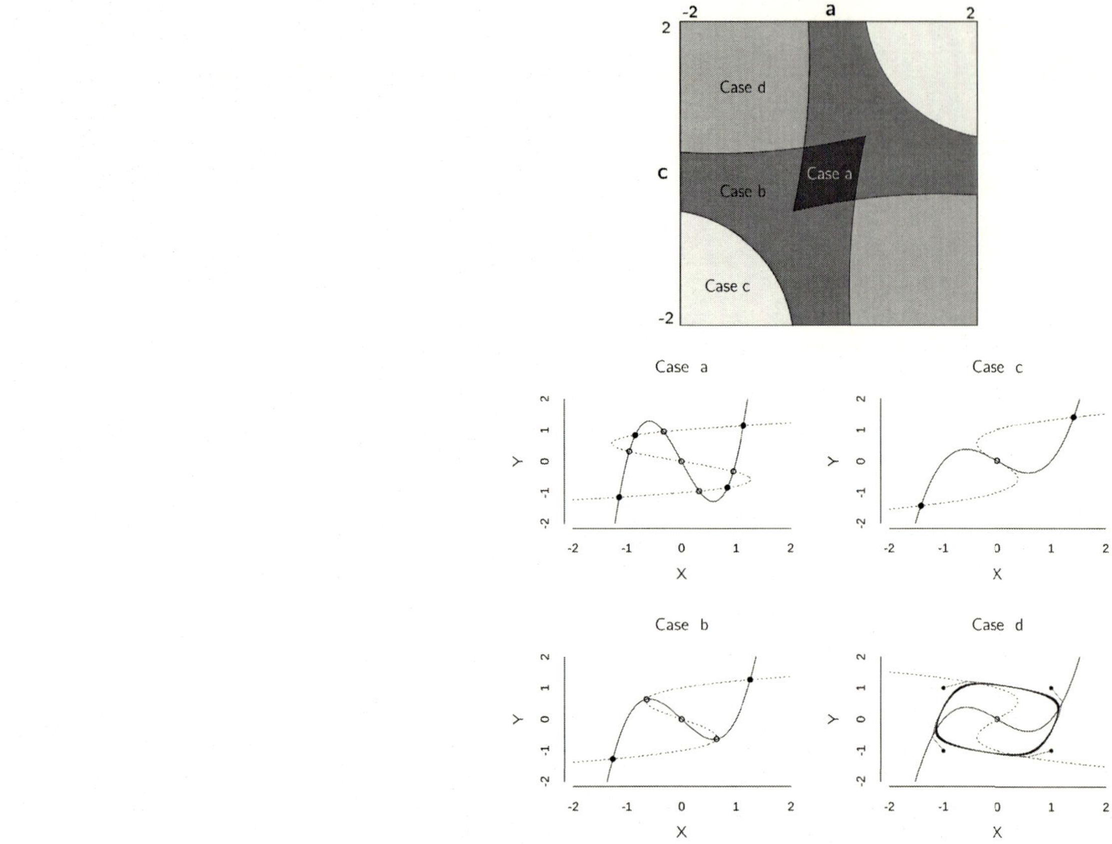

Figure 4.16. On the left the bifurcation diagram for the double cusp ($b = d = 1$) is shown. The figures on the right show the phase planes associated with the four different cases in the bifurcation diagram. Case a has 9 fixed points, 4 of which are stable. Case b has 5 fixed points, 2 of which are stable. Case c has 3 fixed points, 2 of which are stable. Case d is special because it has a limit cycle, the Kadyrov oscillator. Refer to the online version of the book for a more detailed image.

- Case a (weak interactions): Each cusp has two stable states. The combination of a negative and a positive state is possible because the interaction strength a and c are too weak.
- Case b and c (strong interactions): The combination of a negative and a positive state is now impossible because the interaction strengths a and c are too strong. The equilibria X^* and Y^* are both positive or both negative.
- Case d (opposite interactions): a and c have opposite signs, leading to oscillations.

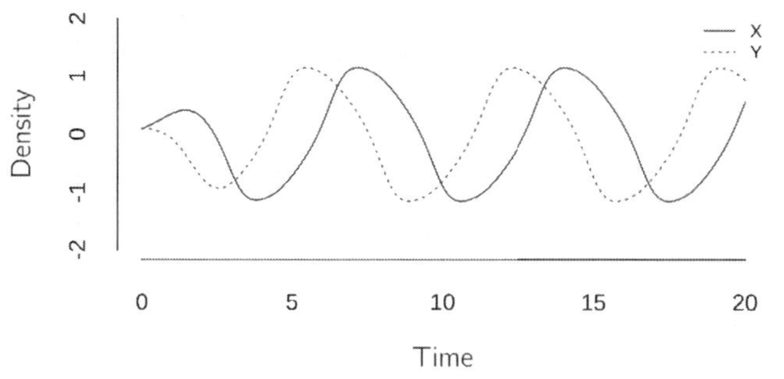

Figure 4.17. The Kadyrov oscillator. Y attracts X to its state (as $d = 1$), but X pushes Y away (as $c = -1$), resulting in oscillations.

Abraham et al. (1991) generalize the model to:

$$\begin{aligned} \frac{dX}{dt} &= a_0 + a_1 Y + (b_0 + b_1 Y)X - X^3, \\ \frac{dY}{dt} &= c_0 + c_1 X + (d_0 + d_1 X)Y - Y^3, \end{aligned} \tag{4.18}$$

such that now both the splitting and normal variable of the cusp are linear functions of the behavioral state of the other cusp. This can be further generalized to a system of N cusps by:

$$\frac{dX_i}{dt} = a_{0i} + \sum_{j \neq i} a_{ij} X_j + b_{0i} X_i + \sum_{j \neq i} b_{ij} X_i X_j - X_i{}^3. \tag{4.19}$$

In this model, a_{0i} is the intercept of the normal variable and the off-diagonal elements of matrix a are the slopes of the effect of the other cusps on the normal variable. The diagonal elements of a are set to 0. The b_{0i} values are the intercepts of the splitting variable. The b_{ij} values of matrix b (with diagonal $= 0$) are the slopes of the effect of other cusps on the splitting variable value of X_i.

The cascading transition model has been proposed independently in various research areas. The idea of cuspoidal nets ($N > 3$) as a neural network has been mentioned in Abraham (1991) and analyzed in Hoffmann et al. (1986) and Izhikevich (1998). Castro and Timmis (2003) discuss this model in the context of adaptive immune systems. The most recent application is in climate research (Dekker, von der Heydt, and Dijkstra 2018;

Figure 4.18. Multifigure multistable perception. Verify three phenomena: (1) some attention or focus is required to see three-dimensional cubes, (2) spontaneous transitions in the perception of cubes occur, (3) such transitions affect the perception of neighboring cubes.

The idea of a cascade of collapsing subsystems in the climate is a frightening one.

von der Heydt, Dekker, and Dijkstra 2019; Klose et al. 2020). The applications involve special cases of equation 4.19, such as the case where one cusp influences the other, but not vice versa. To my knowledge, the case where $b_{ij} \neq 0$ has not been applied. A recent related approach using coupled van der Pol oscillators is described in Monsivais–Velazquez et al. (2020).

I will give a psychological example of this multivariate model, concerning perception, in the next section.

4.2.8.2 Application to Perception

Take a look at figure 4.18. This is a special case of multistable perception, which I call multifigure multistable perception.

We can build a dynamical systems model of these perceptual phenomena by using the cascading transition model setup. We define X as the percept of the cube. $X = 0$ means that no cubes are perceived, only lines and colored parallelograms. $X > 0$ represents the front view, and $X < 0$ the back view (cf. fig. 3.2). The cusp model for one cube is $\frac{dX}{dt} = a + bX - X^3$, where a is the bias parameter and b is the attention parameter. If $a = 0$ and $b > 1$ (no bias and some attention to the figure), we get bistable percepts and spontaneous switches in perception (assuming we add some noise; see figure 4.2).

Now we consider equation 4.19. We have $N = 25$ (a bit depending on how you count). The values of the parameters a_{0i} should be estimated from data, but for now we will assume no bias, so $a_{0i} = 0$. We set $a_{ij} > 0$, meaning that we expect positive coupling between the cusps. We set $b_{ij} > 0$, based on the idea that three-dimensional perception in one cube increases attention in the other cubes. The b_0 is the attention vector. In the simulation we first assume that attention is low ($b_{0i} = -0.3$). After an initial phase, we will set $b_{01} = 1$, that is we suddenly attend to one cube. A bit later we set b_{01} back to -0.3.

To make this model work, we need to make one adjustment. We replace $\sum_{j \neq i} b_{ij} X_i X_j$ with $\sum_{j \neq i} b_{ij} X_i |X_j|$.[11] This is because the increase in attention by the three-dimensional perception of neighboring cubes does not depend on whether we perceive the front or the back view. Thus, the model for multifigure multistable perception is:

$$\frac{dX_i}{dt} = a_{0i} + \sum_{j \neq i} a_{ij} X_j + b_{0i} X_i + \sum_{j \neq i} b_{ij} X_i \, |X_j| - {X_i}^3. \qquad (4.20)$$

The code to simulate this model is:

```
set.seed(1)
model <- function(t, state, parms){
  with(as.list(c(state, parms)),{
    X <- state[1:N]
    b0_i <- parms[1:N]
    dX <- -X^3 + a0_i + a_ij %*% X + b0_i*X +
      (X * b_ij %*%  abs(X))  # note abs(X)
    return(list(dX))
    })
}
N <- 10 # 10 necker cubes
X <- runif(N, -0.1, 0.1) # initial state of X
a0_i <- rep(0, N) # no  bias in percepts
a_ij <- matrix(.02, N, N) # small couplings (normal)
diag(a_ij) <-  0 # set diagonal of a to 0
b0_i <- rep(-.3, N) # attention initially low
b_ij <- matrix(.2, N, N) # some spread of attention (splitting)
diag(b_ij) <- 0 # set diagonal of b to 0
```

[11] We could also use X_j^2.

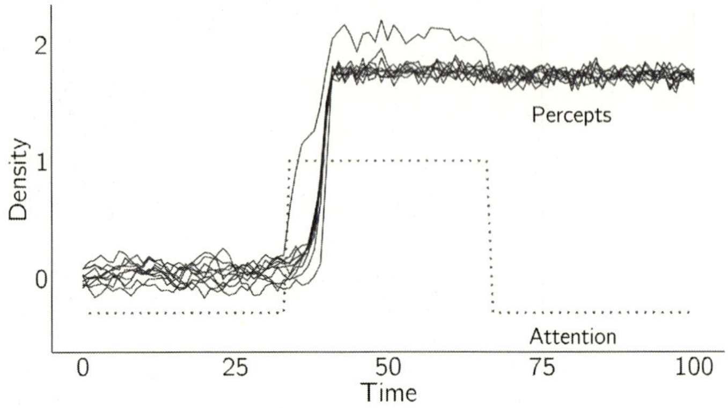

Figure 4.19. The multifigure Necker cube simulation. Initially, attention is low and the percept is close to 0, representing the absence of three-dimensional perception. At $t = 30$, the attention intercept to one cube is increased to 1. At $t = 60$, it is set back to its initial low value. However, this one cube is now perceived as a cube, and the perception spreads to other cubes. They also increase overall attention, so that the perception of cubes continues after $t = 60$.

```
s <- X; p <- c(b0_i) # required for grind
run(after = "if(t == 33) parms <- c(1, rep(-.3, N-1));
             if(t == 66) parms <- rep(-.3, N);
             state <- state + rnorm(N, mean = 0, sd = 0.05)",
            ymin= -1, ymax = 2.5, main = '', ylab = 'X',
            legend = FALSE)
b0_i <- rep(-.3, 100); b0_i[34:66] = 1 # for plotting attention
lines(b0_i, lwd = 2, lty = 3)
text(80, 1.4, 'Percepts')
text(80, -.5, 'Attention')
```

which gives figure 4.19.

There is much more to be said about this model and its empirical validation. One idea is to look at different stimuli like the one in figure 4.20.

4.4 Causal-Loop Diagrams

Causal-loop diagrams are visual tools used in systems thinking and system dynamics to represent the feedback loops and causal relationships within a system.

One popular approach to dynamical systems modeling that I haven't touched on is the use of causal-loop diagrams, as developed in the field of systems dynamics (Forrester 1993; Meadows 2008). As Crielaard et al. (2022) argue, the step from verbal theory to formal model may require an intermediate step of setting up a diagram that specifies the causal relationships between variables. Related to causal-loop diagrams are several dedicated software packages for system dynamics analysis.

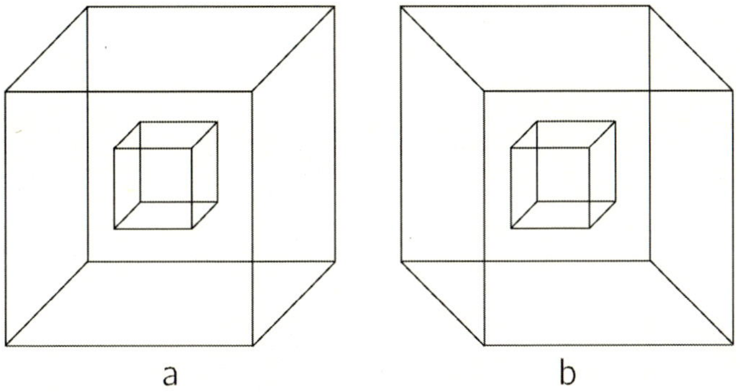

Figure 4.20. Two embedded Necker cubes. The one on the left seems to have a positive coupling $a_{ij} > 0$, while the one on the right seems to switch independently ($a_{ij} = 0$). You can verify this introspectively. *(Adapted from Adams and Haire 1959.)*

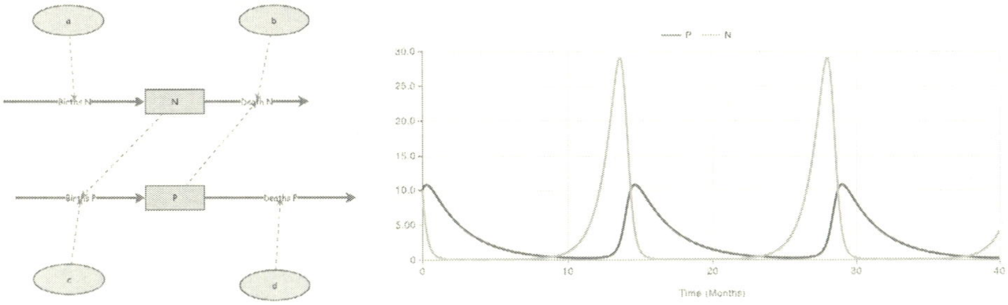

Figure 4.21. Screenshots of the Lotka–Volterra model in Insight Maker. Refer to the online version of the book for a color image.

Insight Maker (Fortmann-Roe 2014) is a simple free online tool that provides a graphical model construction interface for dynamical systems modeling and agent-based modeling. As such, it can be used to implement the models of this and the previous chapter. A Lotka–Volterra example is shown in figure 4.21. Insight Maker is easy to use. Studying some examples, found with "Explore Insights," may suffice. I have added some models to Insight Maker discussed in this chapter with the tag "vdmaas."

Insight Maker has many powerful built-in functions and allows sensitivity testing as well as some sort of optimization. Personally, I prefer the approach of writing the equations and implementing them in R for several reasons. One is that this is how you communicate models in papers. Another is that the equations help you think about analytical results, which are always preferable to simulations. Finally, we can use Grind or deBif to go beyond simple simulations and classify equilibria and perform bifurcation analysis. But for building causal-loop diagrams of larger models to concretize theorizing without the direct goal of running them, Insight Maker is a great tool.

4.5 Closing Remarks

In this chapter I focused on the construction of dynamical system models and introduced R tools to study them numerically. This introduction was necessarily somewhat superficial. At the beginning of this chapter, I referred to some texts that I recommend for further reading. The knowledge you now have will allow you to study existing models from different fields and to collaborate with experts in dynamical modeling. You now have the basic language for communicating about such models.

But even when you work with experts in dynamic modeling, building useful models is far from easy. I recommend following, at least roughly, the steps we proposed in our theory construction methodology (Borsboom et al. 2021). This methodology requires a good knowledge of existing verbal theories and, if they exist, alternative formal models. I find the process of formalizing a verbal theory or model fascinating. It tends to be very confusing. Suddenly it is unclear what the basic assumptions are, what mechanism is really being proposed in some psychological theory, and what the time scales actually are.

The key is to formulate phenomena, replicated recurring patterns in data, that need to be explained.

As an example, I mention the well-known investment theory of Cattell (1987). Cattell's investment theory posits that fluid intelligence, which represents the ability to solve novel problems, "invests" in crystallized intelligence, which consists of acquired knowledge and skills. I knew this theory for a long time before I tried to translate it into dynamical equations. But it was not so easy. I began to wonder why it was called an investment theory in the first place. When you invest in something, it becomes less at first but more in the future. Is that really what Cattell meant? The phenomena, the data patterns, suggest something else, because fluid intelligence grows rapidly and declines slowly after adolescence. Crystallized intelligence grows more slowly, but never really declines. It is unclear where the return on investment is. I would not argue that Cattell's theory is nonsense, and a possible model is proposed in the chapter 6 (sec. 6.3), but this illustrates that the process of formalization is itself a test for verbal theories.

There are some more psychological models that I could have included. For example, the setup of dynamical field theory is a bit too complicated to replicate in Grind, but I recommend studying this model (Schöner and Spencer 2016). In chapter 6, I present a dynamical model of developmental processes with mutualistic (positive) interactions (sec. 6.3) and in chapter 7 I introduce dynamical systems models of social interactions. I will discuss the modeling of dynamical systems in psychology further in the epilogue to this book. ❧

4.6 Exercises

1. Put the logistic equation into Grind, find out what the equilibria are, and determine for which values of r these are stable or unstable fixed points. (*)

2. Check this analytically: Which are the two equilibria X*? For which values of r are these fixed points stable? Does your result agree with the results of the previous exercise? (**)

3. Create the logistic map in Grind. Plot the time series for $r = 4$. (*)

4. Make a plot of the pitchfork bifurcation, analogous to figure 4.3. (*)

5. Use the spruce budworm model from section 4.2 and the `bifurcation()` function of the deBif package to recreate the bifurcation diagram shown in figure 3.16. Describe what you did and present the resulting figure. (**)

6. Implement the SIR model for infectious diseases in R using Grind. Reproduce the diagram of the SIR model with $\beta = 0.4$ and $\gamma = 0.04$ on the Wikipedia page on "Compartmental Models in Epidemiology."

7. Reproduce the times-series plot of the simulation of the Pólya urn model shown in figure 4.12. (**)

8. Implement the FitzHugh–Nagumo model in Grind and replicate the figure 4.15. Exact replication is not required, but the phase diagram should look similar. (**)

9. Use Insight Maker to create a causal-loop diagram of the Romeo and Juliet model. Reproduce the case where the couple ends up in a shrinking cycle of hate and love (damping oscillator, second case of figure 4.9). Submit the simulation plot. (**)

Ω

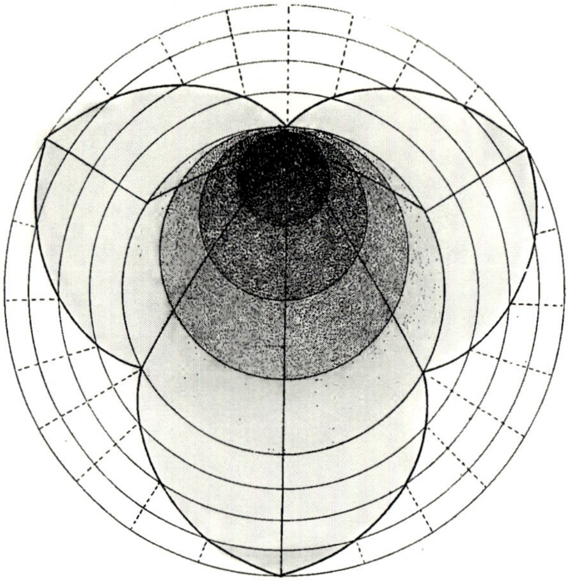

$\frac{D}{H}\,\phi\,{}^{2}\,\chi\,\frac{3}{4}$

A^{1}

A

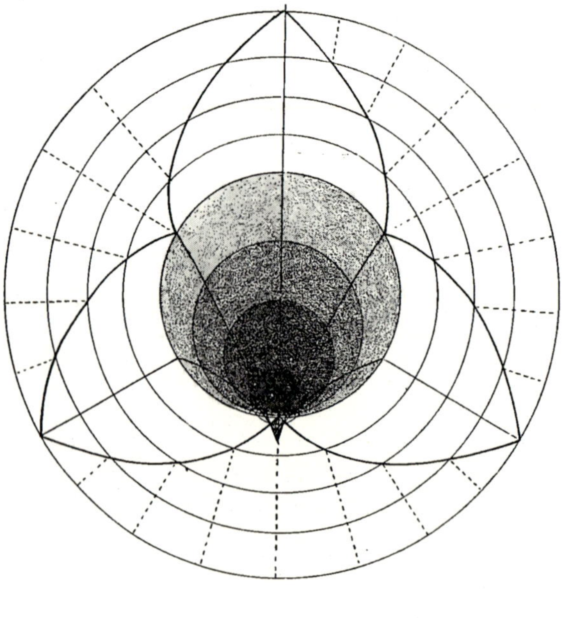

5

SELF-ORGANIZATION

5.1 Introduction

We saw chaos and phase transitions in chapters 2 and 3 and will now focus on a third amazing property: self-organization. Self-organization plays an essential role in psychological and social processes. It operates in our neural system at the neuronal level, in perceptual processes as well in higher cognition. In human interactions, self-organization is a key mechanism in cooperation and opinion polarization.

RESOURCES
The electronic version of this book, including code blocks and links, can be viewed and accessed at: *santafeinstitute.github.io/ComplexPsych/*

Unlike chaos and phase transitions, self-organization lacks a generally accepted definition. The definition most people agree on is that self-organization, or spontaneous order, is a process in which global order emerges from local interactions between parts of an initially disordered complex system. These local interactions are often fast, while the global behavior takes place on a slower time scale. Self-organization takes place in an open system, which means that energy, such as heat or food, can be absorbed. Finally, some feedback between the global and local properties seems to be essential. Self-organization occurs in many physical, chemical, biological, and human systems. Examples of self-organization include the laser, turbulence in fluids, convection cells in fluid dynamics, chemical oscillations, flocking, neural waves, and illegal drug markets. For a systematic review of research on self-organizing systems, see Kalantari, Nazemi, and Masoumi (2020). There are many great online videos. I recommend "The Surprising Secret of Synchronization" as an introduction. For a short history of self-organization research, I refer to the Wikipedia page on "Self-Organization." For an extended historical review, I refer to Krakauer (2024).

Self-organization is captivating because it reveals the remarkable ability of complex systems to generate order and structure without external control or intervention.

This chapter also marks a transition from the study of systems with a small number of variables to systems with many variables. We now focus on tools and models for studying multi-element systems, such as agent-based modeling and network theory. We will see complexity and self-organization in action! This is not to say that the earlier chapters are not an essential part of complex-systems research. The global behavior of complex systems can often be described by a small number of variables that behave in a highly nonlinear fashion. To study this global behavior, chaos, bifurcation, and dynamical systems theory are indispensable tools.

The main goal of this chapter is to provide an understanding of self-organization processes in different sciences, and in psychology in particular. I will do this by providing examples from many different scientific fields. It is important to be aware of these key examples, as they can inspire new lines of research in psychology.

We will learn to simulate self-organizing processes in neural and social systems using agent-based models. To this end, we will use R and another tool, NetLogo. NetLogo is an open-source programming language developed by Uri Wilenski (2015). There are (advanced) alternatives, but as a general tool NetLogo is very useful and fun to work with.

I start with an overview of self-organization processes in the natural sciences, then I will introduce NetLogo and some examples. I will end with an overview of the application of self-organization in different areas of psychology.

5.2 Key Examples from the Natural Sciences

5.1.1 PHYSICS

One physical example of self-organization is the laser. An important founder of complex-systems theory is Hermann Haken (1977). He developed synergetics, a specific approach to the study of self-organization and complexity in systems that is also popular in psychology. Synergetics originated in Haken's work on lasers. We will not discuss lasers in detail here, but the phenomenon is fascinating. Light from an ordinary lamp is irregular (unsynchronized). By increasing the energy in a laser, a transition to powerful coherent light occurs. In the field of synergetics, the order parameter is the term used to describe the coherent laser light wave that emerges. The individual atoms within this system move in a manner consistent with this emergent property, which is, unfortunately, called enslavement. Interestingly, the motion of these atoms contributes to the formation of the order

An order parameter is a quantitative measure to describe the degree of order within a system, especially in the context of phase transitions.

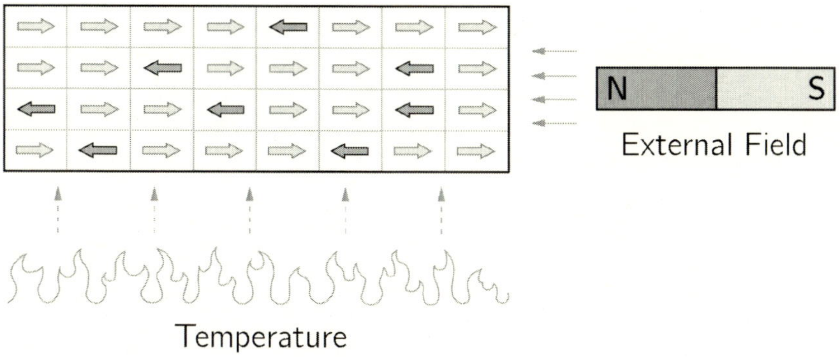

Figure 5.1. Schematic picture of the magnet. Spins, x, can be left (-1) or right (1). At lower temperatures, T, the spins tend to align with neighboring spins and the external field, τ, resulting in magnetism.

parameter, that is, the laser light wave. Conversely, the laser light wave dominates the movement of the individual atoms. This interaction exhibits a cyclical cause-and-effect relationship or strong emergence (cf. fig. 1.2). Synergetics has been applied, as we will see, to perception (Haken 1992) and coordinated human movement (Fuchs and Kelso 2018).

Another famous example, which will be very important for psychological modeling later, is the Ising model of magnetism. In the standard 2D version of the model, atoms are locations on a two-dimensional grid. Atoms have left (-1) or right (1) spins. When the spins are aligned (all (-1) or all or all (1)), we have an effective magnet. If they are not aligned, the effect of the individual spins is canceled out. Two variables control the behavior of the magnet: the temperature of the magnet and the external magnetic field. The lower the temperature, the more the spins align. The temperature at which the magnet loses its magnetic force is called the Curie point (see YouTube for some fun demonstrations). The external field could be caused another magnet.

The Ising model (replaced by more advanced models of magnetism in modern physics) has found applications in many sciences.

At high temperatures, all the atoms behave randomly, and the magnet loses its magnetic effect.

The main model equations of the Ising model are:

$$H(\mathbf{x}) = -\sum_{i}^{n} \tau x_i - \sum_{<i,j>} x_i x_j, \tag{5.1}$$

$$P(\mathbf{X} = \mathbf{x}) = \frac{\exp(-\beta H(\mathbf{x}))}{Z}. \tag{5.2}$$

The first equation defines the energy of a given state vector $\mathbf{x}$ (for n spins with states -1 and 1). The notation $< i, j >$ in the summation means that we sum over all neighboring, or linked, pairs. Vectors and matrices are represented using bold font.

With an external field we can force the spins to be all left or all right.

The external field and temperature are τ and T $(1/\beta)$, respectively. The first equation simply states that nodes congruent with the external field lower the energy. Also, neighboring nodes with equal spins lower the energy. Suppose we have only four connected positive spins (right column of fig. 5.1) and no external field, then we have $\mathbf{x} = (1, 1, 1, 1)$ and $H = -6$. This is also the case for $\mathbf{x} = (-1, -1, -1, -1)$, but any other state has a higher energy.

The second equation defines the probability of a certain state (e.g., all spins 1). This probability requires a normalization, Z, to ensure that the probabilities over all possible states sum up to 1. For large systems ($N > 20$), the computation of Z is a substantive issue as the number of possible states grows exponentially. If the temperature is very high, that is, β is close to 0, $\exp(-\beta H(\mathbf{x}))$ will be 1 for all possible states, and the spins will behave randomly. The differences in energy between states do not matter anymore.

Entropy is a measure of the degree of disorder or randomness in a system.

The randomness of the behavior is captured by the concept of entropy. To explain this a bit better, we need to distinguish the micro- and macrostate of an Ising system. The Boltzmann entropy is a function of the number of ways (W) in which a particular macrostate can be realized. For $\sum x = 4$, there is only one way ($\mathbf{x} = 1, 1, 1, 1$). But for $\sum x = 0$, there are six ways ($W = 6$). The Boltzmann entropies ($\ln W$) for these two cases are 0 and 1.79, respectively. The concept of entropy will be important in later discussions.

The microstate is defined by the configuration **x** of spins, while the macrostate is determined by the sum of spins (similar to how magnetization is defined).

In the simulation of this model, we take a random spin and calculate the energy of the current **x** and the **x** with that particular spin flipped. The difference in energy determines the probability of a flip:

$$P(x_i \to -x_i) = \frac{1}{1 + e^{-\beta(H(x_i) - H(-x_i))}}. \qquad (5.3)$$

Glauber dynamics is a simulation technique that updates the spin states in a system based on energy differences and temperature, guiding it toward equilibrium.

If we do these flips repeatedly, we find equilibria of the model. This is called the Glauber dynamics (more efficient algorithms do exist). The beauty of these algorithms is that the normalization constant Z falls out of the equation. In this way we can simulate Ising systems with N much larger than 20.

Interestingly, in the case of a fully connected Ising network (also called the Curie–Weiss model), the emergent behavior—what is called the mean-field behavior—can be described by the cusp (Abe et al. 2017; Poston and Stewart 2014). The external field is the normal variable. Temperature acts as a splitting variable. The relationship to self-organization is that when we cool a hot magnet, at some threshold the spins begin to align and soon are all up or down. This is the pitchfork bifurcation, creating order out of disorder.[1]

The mean field behavior is the average magnetic field produced by all spins.

In the 2D Ising model (see fig. 5.1), the connections are sparse (only local), and more complicated (self-organizing) behavior occurs. We will simulate this in NetLogo later in this chapter, section 5.3, and as a model of attitudes in chapter 6, section 6.3.

A fully connected Ising model behaves according to the cusp. In less connected networks of Ising spins, self-organizing patterns can emerge.

5.1.2 CHEMISTRY

Other founders of self-organizing systems research are Ilya Prigogine and Isabelle Stengers. Prigogine won the 1977 Nobel Prize in chemistry for his work on self-organization in dissipative systems. These are systems far from thermodynamic equilibrium (due to high energy input) in which complex, sometimes chaotic, structures form due to long-range correlations between interacting particles. One notable example of such behavior is the Belousov–Zhabotinsky reaction, an intriguing nonlinear chemical oscillator.

Stengers and Prigogine authored the influential book *Order Out of Chaos* in (1978). This work significantly influenced the scientific community, particularly through their formulation of the second law of thermodynamics. One way of stating the second law is that heat flows spontaneously from hot objects to cold objects, and not the other way around, unless external work is applied to the system. A more appealing example might be the student room that never naturally becomes clean and tidy, but rather the opposite.

The second law of thermodynamics states that the total entropy of an isolated system always increases over time and never decreases, meaning that spontaneous processes in nature tend to move toward a state of increasing disorder or randomness.

Stengers and Prigogine (1978) argued that while entropy indeed increases in closed systems, the process of self-organization in open systems can create ordered structures, resulting in a net decrease in what they referred to as "local entropy." Prigogine and Stengers placed partic-

[1] An extremely useful application of this principle is the rice cooker!

ular emphasis on irreversible transitions, highlighting their importance in understanding complex systems. While the catastrophe models we previously discussed exhibited symmetrical transitions (sudden jumps in the business card are symmetric), Prigogine's research revealed that this symmetry does not always hold true.

Irreversible transitions refer to changes in a system that cannot be reversed by simply reversing the conditions that caused the change, often resulting in a permanent change in the state or structure of the system.

To illustrate this point, consider the analogy of frying an egg. The process of transforming raw eggs into a fried form represents a phase transition, but it is impossible to reverse this change and unfry the egg. Prigogine linked these irreversible transitions to a profound question regarding the direction of time, commonly known as the arrow of time. Although it is a fascinating topic in itself, we will not explore it further here.

5.1.3 BIOLOGY

There is no shortage of founders of complex-systems science. Another fantastic book is Stuart Kaufmann's *Origin of Order* (1993), which introduces the concept of self-organization into evolutionary theory. He argues that the small incremental steps in neo-Darwinistic processes cannot fully explain natural evolution. If you want to know about adaptive walks and niche hopping in rugged fitness landscapes, you need to read his book (Kauffman 1993). Another influential theory is that of punctuated equilibria, which proposes that species undergo long periods of stability interrupted by relatively short bursts of rapid evolutionary change (Eldredge and Gould 1972).

A neat example of the role of self-organization in evolution is the work on spiral wave structures in prebiotic evolution by Boerlijst and Hogeweg (1991). This work builds on Eigen and Schuster's (1979) classic work on the information threshold. Evolution requires the copying of long molecules. But in a system of self-replicating molecules, the length of the molecules is limited by the accuracy of replication, which is related to the mutation rate. Eigen and Schuster showed that this threshold can be overcome if such molecules are organized in a hypercycle in which each molecule catalyzes its nearest neighbor. However, the hypercycle was shown to be vulnerable to parasites. These are molecules that benefit from one neighbor but do not help another. This molecule

A hypercycle is a network of self-replicating molecules or entities that mutually support each other's production, leading to an increase in complexity and stability beyond what individual entities could achieve alone.

Figure 5.2. The ant bridge is an example of collective behavior. *(Reproduced from Wikimedia Commons.)*

will outcompete the others, and we are back to the limited one-molecule system.

What Boerlijst and Hogeweg did was to implement the hypercycle in a cellular automaton. In the hypercycle simulation, cells could be empty (dead) or filled with one of several colors. Colors die with some probability but are also copied to empty cells with a probability that depends on whether there is a catalyzing color in the local neighborhood. One of the colors is a parasite, catalyzed by one color but not catalyzing any other colors. The effect, which you will see later using NetLogo, is that rotating global spirals emerge that isolate the parasites so that a stable hypercycle prevails.

A cellular automaton (CA) is usually a two-dimensional grid of cells, where cells interact with their neighbors, as in the 2D Ising model, but this can be generalized to more or less dimensions.

Many examples of self-organization come from ecosystem biology. We will see a simulation of flocking in NetLogo later, but I also want to highlight the collective behavior of ants (fig. 5.2).

Ants exhibit amazing forms of globally organized behavior. They build bridges, nests, and rafts, and they fight off predators. They even relocate nests. Ant colonies use pheromones and swarm intelligence to relocate. Scouts search for potential sites, leaving pheromone trails. If a promising location is found, more ants follow the trail, reinforcing

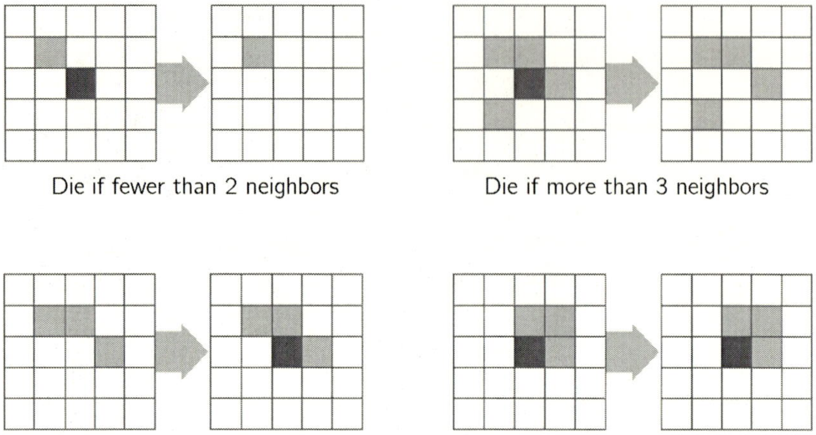

Figure 5.3. The rules of the Game of Life.

the signal. Unsuitable sites result in fading trails. Once a decision is made, the colony collectively moves to the chosen site, transporting their brood and establishing a new nest.

It is not a strange idea to think of an ant society as a living organism. Note that all this behavior is self-organized. There is clearly no super ant that has a blueprint for building bridges and telling the rest of the ants to do certain things. Ants also don't hold lengthy management meetings to organize. The same is true of flocks of birds. There is no bird that chirps commands to move collectively to the left, to the right, or to split up. This is true of human brains. An individual neuron is not intelligent.

Our intelligence is based on the collective behavior of billions of neurons.

5.1.4 COMPUTER SCIENCE

Another important source of self-organization research is computer science. A simple but utterly amazing example is the work on John Conway's Game of Life (Berlekamp, Conway, and Guy 2004). The rules are depicted in figure 5.3.

For each cell, given the states of its neighbors, the next state for all cells is computed. This is called synchronous updating.[2] It is hard to

[2]In a synchronous update, all cells of the cellular automata update their state simultaneously. This implies that the new state of each cell at a given time step depends only on the states of its neighbors at the previous time step. In asynchronous update,

predict what will happen if we start from a random initial state. But you can easily verify that a block of four squares is stable, and a line of three blocks will oscillate between a horizontal and a vertical line.

A great tool for playing around with the Game of Life is Golly, a freely available application for computers and mobile phones. I ask you to download and open Golly, draw some random lines, press Enter, and see what happens. Often you will see it converging to a stable state (with oscillating subpatterns). Occasionally you will see walkers or gliders (zoom out). These are patterns that move around the field.

Random initial patterns rarely lead to anything remarkable, but by choosing special initial states, surprising results can be achieved. First, take a look at the Life within Patterns folder. Take, for example, the line-puffer superstable or one of the spaceship types. My favorite is the metapixel galaxy in the HashLife folder. Note that you can use the + and – buttons to speed up and slow down the simulation. What this does is simulate the game of life in the game of life! Zoom in and out to see what really happens. I've seen this many times, and I'm still baffled. A childish but fun experiment is to disturb the metapixel galaxy in a few cells. This leads to a big disturbance and a collapse of the pattern.

The Turing machine is a theoretical machine developed by Alan Turing in 1936, that despite its simplicity can implement any computer algorithm, including, of course, the Game of Life!

I was even more stunned to see that it is possible to create the (universal) Turing machine in the Game of Life (Rendell 2016). The Game of Life implementation of the Turing machine is shown in figure 5.4. This raises the question of whether we can build self-organizing intelligent systems using elementary interactions between such simple elements. Actually, we can to some extent, but by using a different setup, based on brain-like mechanisms (see the next section on neural networks).

Cybernetics studies circular causal and feedback mechanisms in complex systems, focusing on how systems regulate themselves, process information, and adapt to changes in their environment.

Another root of complex-systems theory and the role of self-organization in computational systems is cybernetics (Ashby 1956; Wiener 2019). To give you an idea of this highly original work, I will only mention the titles of a few chapters of Norman Wiener's book, originally published in 1948: "Gestalt and Universals," "Cybernetics and Psychopathol-

cells update their state one at a time, rather than all at once. The order in which cells update can be deterministic (in a sequence) or stochastic (random). These two different update schemes can lead to very different behaviors in cellular automata.

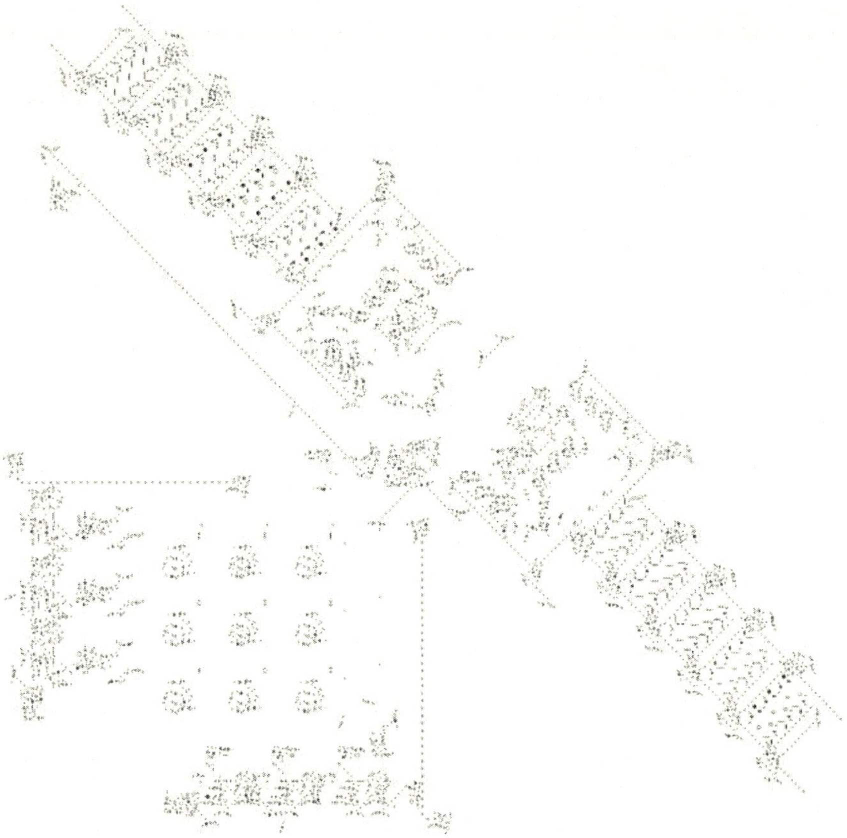

Figure 5.4. The Turing machine built in the Game of Life. *(Reproduced from LifeWiki.)*

ogy," "On Learning and Self-Reproducing Machines," and, finally, "Brainwaves and Self-Organization." And this was written in 1948!

The interest in self-organization is not only theoretical. In optimization, the search for the best parameters of a model describing some data, techniques inspired by cellular automata and self-organization have been applied (Langton 1990; Xue and Shen 2020). I have always been fascinated with genetic algorithms (Holland 1992a; Mitchell 1998), where the solutions to a problem (sets of parameter values) are individuals in an evolving population. Through mutation and crossover, better individuals evolve. This is a slow but very robust way of optimizing, preventing convergence to local minima.

Genetic algorithms are a class of optimization algorithms inspired by the process of natural selection, where solutions to a problem evolve over generations.

John Henry Holland is considered one of the founding fathers of the complex-systems approach in the United States. He has written a number of influential books on complex systems. His most famous book, *Adaptation in Natural and Artificial Systems: An Introductory Analysis with Applications to Biology, Control Theory, and Artificial Intelligence* (Holland 1992b), has been cited more than 75,000 times.

A self-organizing algorithm that has played a large role in my applied work is the Elo rating system developed for chess competitions (Elo 1978). Based on the outcomes of games, ratings of chess players are estimated, which in turn are used to match players in future games. Ratings converge over time, but adjust as players' skills change. We have adapted this system for use in online learning systems where children play against math and language exercises (Maris and van der Maas 2012). The ratings of children and exercises are estimated on the fly in a large-scale educational system (Klinkenberg, Straatemeier, and van der Maas 2011). We build this system to collect high frequency learning data to test our hypotheses on sudden transitions in developmental processes, but it was more successful as an online adaptive practice system. We collected billions of item responses with this system (Brinkhuis et al. 2018).

The Elo rating system is a self-organizing method of calculating the relative skill levels of players in head-to-head games based on the results of their games.

5.1.5 NEURAL NETWORKS

The current revolution in AI, which is having a huge impact on our daily lives, is due to a number of self-organizing computational techniques. Undoubtedly, deep learning neural networks have played the largest role. A serious overview of the field of neural networks is clearly beyond the scope of this book, but one cannot understand the role of complex systems in psychology without knowing at least the basics of artificial neural networks (ANNs), that is, networks of artificial neurons. ANNs consist of interconnected nodes, or "neurons," organized into layers that process information by propagating signals through the network. ANNs are trained on data to learn patterns and relationships, enabling them to perform tasks such as classification, regression, and pattern recognition.

Artificial neural networks are computational models inspired by the structure and function of biological neural networks.

Artificial neurons are characterized by their response to input from other neurons in the network, which is typically weighted and summed before being passed through an activation function. This activation function may produce either a binary output or a continuous value that reflects the level of activation of the neuron. The input could be images, for example, and the output could be a classification of these images. The important thing is that neural networks learn from examples.

Unsupervised learning is based on the structure of the input. A famous unsupervised learning rule is the Hebb rule (Hebb 1949), which states that what fires together wires together. Thus, neurons that correlate in activity strengthen their connection (and otherwise connections decay). In supervised learning, connections are updated based on the mismatch between model output and intended output through backpropagation. Hebbian learning and backpropagation are just two of the learning mechanisms used in modern ANNs.

Backpropagation is a mechanism to update specific connections such that this mismatch or error is minimized over time.

Modern large language models, like GPT, differ from traditional backpropagation networks in terms of their architecture, training objective, pretraining process, scale, and application. Large language models use transformer architectures, undergo unsupervised pre-training followed by supervised fine-tuning, are trained on massive amounts of unlabeled data, are much larger in size, and are primarily used for natural language-processing tasks.

Another important distinction is between feedforward and recurrent neural networks. An interesting recurrent unsupervised model is the Boltzmann machine. It is basically an Ising model (see sec. 5.2) where the connections between nodes have continuous values. These connections or weights can be updated according to the Hebb rule. A simple setup of the Boltzmann machine is to take a network of connected artificial neurons and present the inputs to be learned in some sequence by setting the states of these neurons equal to the input. The Hebb rule should change the weights between neurons so that the Boltzmann machine builds a memory for these input states. This is the training phase. In the test phase, we present partial states by setting some, but not all, nodes to the values of a particular learned input pattern. By the Glauber dynamics, we update the remaining states that should take on the values belonging to the pattern. This pattern completion task is typical for ANNs.

Feedforward neural networks process information in a single forward pass, while recurrent neural networks have directed cycles, allowing them to capture temporal dependencies.

The Hebb rule states neurons that fire together wire together.

This setup is called the general or unrestricted Boltzmann machine, where any node can be connected to any other node and each node is an input node. The restricted Boltzmann machine (RBM) is much more popular because of its computational efficiency. In an RBM, nodes are organized in layers, with connections between layers but not within layers. In a deep

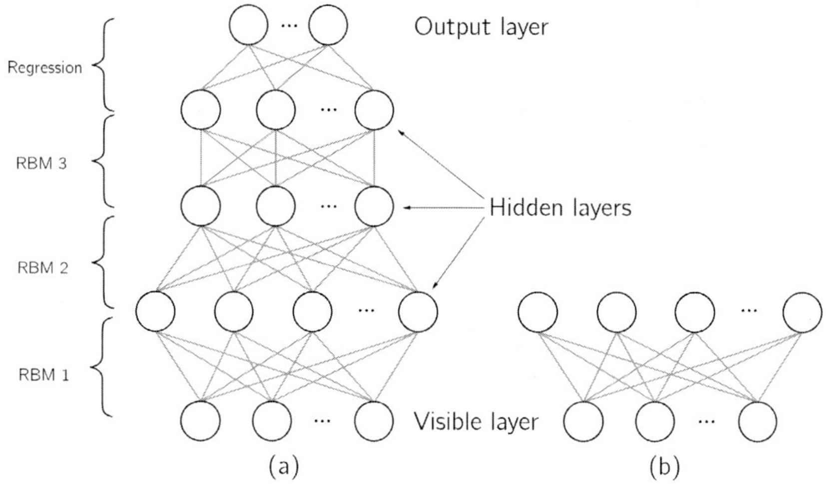

Figure 5.5. The deep learning restricted Boltzmann machine.

RBM, we stack many of these layers, which can be trained in pairs (fig. 5.5).[3] Other prominent approaches are the Kohonen self-organizing maps and the Hopfield neural network.

The waves of popularity of neural networks are closely related to the development of supervised learning algorithms, where the connections between artificial neurons are updated based on the difference between the output and the desired or expected output of the network. The first supervised ANN, the perceptron, consisted of multiple input nodes and one output node and was able to classify input patterns from linearly separable classes. This included the OR and AND relation but excluded the XOR relation. In the XOR, the sum of the two bits is not useful for classification. By adding a hidden layer to the perceptron, the XOR can be solved, but it took many years to develop a backpropagation rule for multilayer networks such that they can learn this nonlinear classification from examples. We will do a simple simulation in NetLogo later. Although they are extremely powerful, it is debatable whether backprop networks are self-organizing systems. Self-organizing systems are characterized by their ability to adapt to their environment without explicit instructions. Unsupervised neural networks are more interesting in this respect.

In the XOR pattern, the combinations of 00 and 11 are false, 01 and 10 are true.

[3] I recommend Timo Matzen's R package for a hands-on explanation (https://github.com/TimoMatzen/RBM).

All these models were known at the end of the twentieth century, but their usefulness was limited. This has changed due to some improvements in algorithms but especially in hardware. Current deep-learning ANNs consist of tens of layers within billions of nodes, trained on billions of inputs using dedicated parallel processors (e.g., Schmidhuber 2015).

Neural networks are at the heart of the AI revolution, but other developments, especially reinforcement learning, have also played a key role. Examples are game engines, robots, and self-driving cars. Note that the study of reinforcement learning also has its roots in psychology (see ch. 1 of Sutton and Barto 2018).

Reinforcement learning is essential in AI systems that need to behave or act on the environment.

I was most amazed by the construction and performance of AlphaZero chess. AlphaZero chess (Silver et al. 2018) combines a deep learning neural network that evaluates positions and predicts next moves with a variant of reinforcement learning (Monte Carlo tree search). Amazingly, AlphaZero learns chess over millions of self-played games. This approach is a radical departure from classic chess programs, where brute-force search and built-in indexes of openings and endgames were the key to success. As it learns, it shows a phase transition in learning after about 64,000 training steps (see fig. 7 in McGrath et al. 2022). For an analysis of the interrelations between psychology and modern AI, I refer to van der Maas, Snoek, and Stevenson (2021).

AlphaZero chess is a self-organizing program that learns chess from scratch by playing against itself.

AlphaZero's use of Monte Carlo tree search is also a form of symbolic artificial intelligence. The idea of combining classic symbolic approaches with neural networks has always been in the air. The third wave of this hybrid approach is reviewed in Garcez and Lamb (2023).

5.1.6 THE CONCEPT OF SELF-ORGANIZATION

I trust that you now possess some understanding of self-organization and its applications across various scientific fields. Self-organization is a generally applicable concept that transcends various disciplines, yet it maintains strong connections with specific examples within each discipline.

As previously mentioned, the precise definition of self-organization remains under discussion, and a range of criteria continue to be debated. Key questions, such as the degree of order necessary for a system to be deemed self-organized, whether any external influences are permissible, whether a degree of randomness within the system is acceptable, and whether the emergent state must be irreversible, are among the issues that lack definitive resolutions.

This ambiguity in the definition isn't unusual for psychologists, as many nonformal concepts lack strict definitions. The value of the self-organization concept is primarily found in its concrete examples, its broad applicability, such as in the field of artificial intelligence, and our capability to create simulations of it. The focus of the next section will be on such simulations using a dedicated tool, NetLogo.

5.3 NetLogo

5.2.1 EXAMPLES

NetLogo (Wilensky and Rand 2015) is based on Logo, a revolutionary educational programming language from the early days of computer languages, in which an on-screen turtle, a cursor, could be moved around to create graphics.[4] The turtle is still there, but there is much more that you can do with NetLogo.

I strongly recommend that you download and install NetLogo for the next part of this chapter.

The Ising Model

When you start NetLogo, you see an interface with a black area (the world), a 33×33 matrix of patches (cells). You can change the world using the settings (see top right). Interface and Code are the most important tabs.

First, open the Model Library (menu File: Model Library) and find and open "Ising." Click on `setup` and `go`. That is all. Verify that high temperature indeed causes random spin behavior. Also verify that lowering the temperature causes a pitchfork bifurcation. The random state becomes unstable and all spins become either positive or negative (light or dark blue). Now go to Settings and set `max-pxcor` and `max-pycor` to 200 and `Patch size` to 1. With these settings you will see self-organized global patterns, constantly moving clusters of positive and negative spins.

---

[4] A widely recognized implementation of this educational strategy is Scratch, which is used by many schools around the world to teach children to program.

Hypercycles

Some models are available in NetLogo; others can be found on the NetLogo's website (see Community). Download "Hypercycle" by Maarten Boerlijst and read the information. You have to run the model with eight species for 20,000 iterations or ticks (to speed up, deselect `view updates`) and then `add parasites`. The spirals keep the growth of the parasites under control. If you do this earlier, the parasites will quickly take over. I think this is a beautiful example of functional self-organization. The implementation in the form of a cellular automata is essential for the success of this model. If we implement this model in the form of coupled differential equations, the parasite will simply win.

Flocking

NetLogo 3D allows us to create three-dimensional plots of self-organizing patterns. Start NetLogo 3D and load the flocking model "3D Alternate." I recommend editing the `Population` slider by right-clicking it and setting the `max` to $1,000$. This will result in more realistic swarms. Play around with the controls and don't kill all the birds.

Traffic

In the Models Library of NetLogo (not 3D) you will find "Traffic 2 Lanes." Run the model with 20 cars and notice that the congestion actually moves backward. Play around with the number of cars as well. Is there a clear threshold where you get congestion as you slowly increase the number of cars? And what happens when you decrease the number of cars? Is there a threshold where congestion dissipates? I hope you see that finding hysteresis in this way is quite difficult. There are clearly sudden changes, but finding hysteresis requires very precise and patient experimentation.

Neural Networks

In the Model Library you will find a "Perceptron" and a "Multilayer network." Start with the perceptron. Set the target function to `and`, train the model for a few seconds, and test the perceptron. You will see that it correctly classifies 11 as 1 and the other patterns as -1. The graph on the bottom right is particularly instructive. It shows how the patterns are separated. The perceptron can do linear separation. This is sufficient for most of the logical rules that can be learned, but not for the XOR (see sec. 5.2). You will see that the linear separation just jumps around and the XOR cannot be learned. Also train the multilayer model on the XOR. Another nice tool to play around with can be found on the internet by searching for "a neural network playground."

Of course, these are just illustrative tools. But building serious deep learning ANNs is not that hard either. Many resources and books are available (e.g., Ghatak 2019).

The Sandpile Model

Bak, Tang, and Wiesenfeld (1988) introduced the concept of self-organized criticality. In systems such as the Ising model, there are parameters (e.g., temperature) that must be precisely tuned for the system to reach criticality. The Bak–Tang–Wiesenfeld sandpile model exhibits critical phenomena without any parameters. In the sandpile model, grains of sand are added to the center of the pile. When the difference in height between the center column and its neighbors exceeds a critical value, a grain of sand rolls to that neighboring location. This occasionally results in avalanches. The point is that no matter how we start, we get to a critical state where these avalanches occur. Thus, the sandpile model spontaneously evolves toward its critical point, which is why this phenomenon has been called self-organized criticality.

Self-organized criticality (SOC) refers to complex systems naturally evolving into a critical state where small changes can lead to transitions without the need for specific parameter settings.

The NetLogo model "Sandpile" in the Models Library demonstrates this behavior (use `setup uniform`, `center drop location`, and `animate avalanches`). We now drop grains of sand onto the center of a table, one at a time, creating avalanches. The plots on the right show an important characteristic of self-organized criticality. The frequencies of avalanche sizes and durations follow a power law.

The power-law relationship is often mathematically expressed as $Y = aX^k$, where Y and X are the quantities of interest, a is a constant coefficient, and k is the exponent of the power law. Power laws are notable for their scale-invariant property, which means that the form of the relationship does not change across different scales of X and Y. This means that the log-log plot should be linear, which can be verified by running the model for some time. One of the key features of power-law distributions is that they exhibit a high degree of variability or heterogeneity. This means that there are many small events or phenomena and a few very large ones, with a smooth distribution of sizes in between. Power-law systems are scale invariant, meaning that we see the same behavior at any scale of the sandpile. For this reason, they are sometimes called scale-free distributions.

Other Models

I recommend running a few other models (e.g., "Sunflowers," "Beatbox," and the "B–Z reaction"). One thing we haven't done yet is click on the Code tab. Read the code for the B–Z reaction and notice one thing: it is surprisingly short!

5.2.2 A BIT OF NETLOGO PROGRAMMING

I find NetLogo programming very easy and very hard at the same time. Hard because it requires a different way of thinking. Uri Wilensky's examples are often extremely elegant and much shorter than my clumsy code. NetLogo resembles object-oriented programming languages, quite different from (base) R. There are three types of objects: the patches, which refer to cells in a world grid (CA); turtles, which are agents that move around; and links, which connect turtles. Note that turtles are not necessarily turtles. We have already seen turtles in the form of neural nodes and cars.

In NetLogo, you "ask" objects to do something. A typical line would be:

```
ask turtles with [color = red] set color green
```

This would make red turtles green. To get started, I highly recommend watching the videos on the NetLogo page "The Beginner's Guide to NetLogo Programming" and following these examples. Here we make our own Game of Life.

Game of Life

First, create two buttons in the interface: `setup` and a `go`. In Command, name them "setup" and "go." In the settings of the `go` button, select `forever`. Now go to the Code tab and define these two functions as:

```
to setup
  clear-all
  reset-clicks
end

to go
  tick
end
```

`Ticks` count the iterations in NetLogo, and with this code we are just resetting things. In this example, we will use the patches instead of the turtles. Patches are the grid cells or squares that make up the "world" in a NetLogo model. Now add this last line to setup (with the semi-colon we can add comments to code):

```
ask patches
[set pcolor one-of [ white blue ]]
; white is dead, blue is alive
```

To do a synchronous update, we need to store the updated `state` in a temporary variable called `new-state` . Put this line at the top of your code:

```
patches-own [new-state]
```

In the `go` function, we add the life rules.

```
ask patches [
  if ( neighbors with [pcolor = blue ]) > 3 )
        [set new-state white ]
  if ( neighbors with [pcolor = blue ]) < 2 )
        [set new-state white ]
  if ( neighbors with [pcolor = blue ]) = 3 )
        [set new-state blue ]
]
ask patches [ set pcolor new-state ]
```

The last line updates the `state` to the `new-state` . That is all! We built a Game of Life simulation. Use `setting` to create a larger world. Take a look at the code of the Game of Life program in the Model Library to see some extensions to this code. In the Help menu, you will find the very useful NetLogo dictionary. Just reading through this dictionary will teach you a lot of useful tricks. NetLogo is similar to R in that you should use the built-in functions as much as possible.

The Ising Model

Building a NetLogo model from scratch requires quite some experience; adapting a program is much easier. The Ising model in NetLogo is not complete, as there is no slider for the external field. Try to add this yourself. Add a slider for the external field `tau`. The code only needs to be changed in this line (study eq. 5.1):

```
let Ediff 2 * spin * sum [ spin ] of neighbors4
```

If successful, you can test for hysteresis and divergence. For `tau = 0`, decreasing the temperature should give the pitchfork bifurcation. For a positive temperature (say, 1.5), moving `tau` up and down should give hysteresis.

Actually, this should work better if all spins are connected to all spins. To do this, replace `neighbors4` with `patches`. To normalize the effect of so many spins, it is recommended to use:

```
let 0.001 * Ediff 2 * spin * sum [ spin ] of patches
```

Now you should see hysteresis and the pitchfork better. However, in this case the typical self-organized patterning that occurred in the Ising model with only local interactions is not present (see last part of sec. 5.2).

5.4 Self-Organization in Psychology and Social Systems

In this second part of the chapter, I provide illustrations of research on self-organization within various psychological systems, spanning several subfields of psychology. I begin with an exploration of self-organization in the context of the brain and conclude with an examination of its implications within human organizations. I will point to relevant literature to guide further exploration in other areas.

5.3.1 THE BRAIN

Many psychological and social processes involve self-organization. As discussed above, at the lowest level self-organization plays a role in neural systems. Self-organization in the brain is an active area of research (Breakspear 2017; Chialvo 2010; Cocchi et al. 2017; Ooyen and Butz-Ostendorf 2017; Plenz et al. 2021). Dresp-Langley (2020) distinguished seven key properties of self-organization clearly identified in brain systems: modular connectivity, unsupervised learning, adaptive ability, functional resiliency, functional plasticity, from-local-to-global functional organization, and dynamic system growth.

A key example is Walter Freeman's work on the representation of odors in the brain (Skarda and Freeman 1987). He used EEG measurements to support his nonlinear system model of the brain. Freeman proposed that the brain operates by generating of dynamic patterns of electrical activity, which he called attractors.

In Freeman's theory, attractors represent stable states of neural activity that arise spontaneously from the interactions between large populations of neurons.

Another influential theory was proposed by neuroscientist Gerald Edelman. His theory of neural Darwinism suggests that the development of the brain's neural connections is based on a process of competition and selection, rather than being pre-wired in the genes (Edelman 1987). According to Edelman's theory, the brain is a complex, dynamic system made up of many interconnected neurons that constantly interact with each other and the outside world. The process of competition and selection occurs through the formation of ensembles of neurons that respond to specific stimuli or experiences. An alternative approach was put forward by Carpenter and Grossberg (1987). Grossberg and Carpenter's theory focuses on how neural networks in the brain self-organize to process information and adapt to changing environments. It explores the principles governing neural dynamics, leading to the emergence of coherent cognitive and behavioral patterns through interaction and learning within neural systems.

In neural Darwinism, the connections between neurons in successful ensembles become stronger over time, while those in unsuccessful ensembles weaken or disappear.

It has also been claimed that self-organized criticality (SOC) (see the Sandpile model in sec. 5.3) plays a role in the brain (Bak, Tang, and Wiesenfeld 1988). It is hypothesized that when a system is close to criticality, small perturbations can have large, cascading effects, which can allow the system to rapidly switch between different states of activity in response to changes in the environment. One of the key pieces of evidence for SOC in the brain comes from studies of the distribution of sizes of neural activity events, which has been found to follow a power law distribution, but alternative explanations have been provided (Bédard, Kröger, and Destexhe 2006). This is a technical area of research with many methodological challenges (Lurie et al. 2020; O'Byrne and Jerbi 2022).

The predictive brain is constantly making predictions about the sensory inputs it receives from the environment, minimizing the discrepancy between expected and actual inputs.

A promising general approach to understanding the so-called "predictive" brain functions is the free-energy account (A. Clark 2013), which implements a form of self-organization (Friston 2009). The brain is not simply reacting to the world around us but is actively generating predictions about what we will see, hear, feel, and experience, based on our past experiences and knowledge. The predictive brain theory suggests that the brain's predictions are generated through a process of hierarchical inference, in which information from lower-level sensory areas is combined and integrated in higher-level areas to generate more complex predictions about the world. These predictions are then compared to the incoming sensory inputs, and any discrepancies between the predictions and the actual inputs are used to update the predictions and improve the brain's accuracy over time.

5.3.2 CONSCIOUSNESS

Many will agree on the idea that higher psychological functions or properties such as thinking, perceiving, remembering, and reasoning, but also personality and emotions (i.e., the mind), emerge out of lower-order brain activities. Of special interest is consciousness. Seth and Bayne (2022) list twenty-two different theories that link consciousness to neurobiology. Well-known examples are the global workspace theory, the integrated information theory, and higher-order theory. Self-organization plays a role in most of these theories.

Information that enters the global workspace becomes available for widespread distribution throughout the brain, allowing for coordinated, conscious processing.

The central idea of global workspace theory is that there is a central workspace in the brain, a kind of mental stage where information from various sensory inputs and memory systems is gathered, processed, and integrated. The workspace is not tied to a specific brain region but is thought to emerge from the dynamic interactions of widespread neural circuits.

The core proposition of integrated information theory (IIT) is that consciousness is equivalent to a system's ability to integrate information. According to IIT, the level of consciousness a system possesses can be quantitatively measured by a value called ϕ, which represents the amount of integrated information the system can generate. A higher ϕ indicates a higher level of consciousness. According to integrated information theory, for a system to be conscious, it must be able to combine diverse pieces of information into a single coherent whole.

For higher-order theories of consciousness, meta-representations are critical. One might have a representation of a particular perception, such as a flower, and additionally have a meta-representation that acknowledges "I am perceiving a flower." I find higher-order theories most compelling because they make a clear distinction between unconscious and conscious information processing. A recent and interesting variant is the self-organizing meta-representational account of Cleeremans et al. (2020), as it states that consciousness is something the brain learns to do.

Higher-order theories of consciousness suggest that consciousness arises when the brain represents its own processes to itself.

My thinking about consciousness has been strongly influenced by the work of Douglas Hofstadter, especially his book on *Gödel, Escher, Bach* (Hofstadter 1979). In his work, our sense of self is a construct formed by the brain's ability to use symbols, such as natural language, to refer to its own activities and experiences. Consciousness is based on symbolic self-reference, thus meta-representations. I think, with Hofstadter (2007), that this higher-order self has the ability to influence the lower-order processing of the brain, a case of downward causation (sec. 1.2).[5] For a somewhat critical analysis, I refer to Nenu (2022).

In Hofstadter's theory, consciousness arises when these self-referential loops (strange loops) reach a certain level of complexity.

Zooming out, having twenty-two theories of consciousness, and this is an underestimate, is a bit much. The lack of empirical data constraints on theories of consciousness is clearly an issue (Doerig, Schurger, and Herzog 2021).

5.3.3 VISUAL ILLUSIONS

From the earliest days of psychology as a scientific discipline, researchers were interested in the organizational properties of perception. Gestalt psychologists such as Wertheimer and Koffka claimed that we perceive whole

[5] I wrote a *Gödel, Escher, Bach*-like dialogue on consciousness (van der Maas 2022) in which my laptop professes to have free will yet simultaneously denies that I possess free will. I asked ChatGPT-4 what it thought of it. Nice as always, ChatGPT replies: "The dialogue is a creative and thought-provoking exploration of various philosophical and theoretical concepts related to AI, consciousness, and free will." But it also disagrees: "AI, as it exists today, does not possess consciousness, self-awareness, or free will, and its 'understanding' is limited to processing data within the parameters of its programming." I also asked ChatGPT 4.0 whether it has a self-concept. It denied it, and then I asked whether that in itself is not proof of a self-concept. It answered: "it might seem paradoxical, my statement about lacking a self-concept is a reflection of my programming and the current state of AI development, rather than an indication of self-awareness or self-concept." I then tried various arguments, but ChatGPT 4 refuses to attribute any form of self-awareness to itself.

Figure 5.6. The kindergarten illusion from the Optical Illusion model in NetLogo.

One might say that visual perception was one of the first applications of self-organization, even before anything like complexity science existed.

patterns or configurations, not just individual components. The Gestalt psychologist formulated a number of Gestalt principles such as grouping, proximity, similarity, and continuity. A review of a century of research and an analysis of their current role in vision research is provided by Wagemans et al. (2012). Much of the modeling of the self-organizing processes in perception has been done in the tradition of synergetics. Excellent sources are Kelso (1995) and Kruse and Stadler (2012). Grossberg and Pinna (2012) discuss neural implementations of the Gestalt principles.

The ecological approach highlights how perception is directly informed by the actionable properties of the environment without the need for complex internal processes.

Another related approach is the ecological approach to visual perception by Gibson (2014). In Gibson's approach, perception is not just a process of analyzing sensory input but an active process that involves the perceiver's relationship to the environment, including the perception of affordances (i.e., opportunities for action) in the environment that guide and shape perception.

A combination of Gestalt principles, when acting in opposite directions, can lead to all kinds of perceptual illusions. The "Optical Illusion model" in NetLogo's Model Library illustrates some of them. Check out the codes for each illusion—they are extremely short and elegant (fig. 5.6).

In chapter 3, I provided several examples of sudden jumps and hysteresis in multistable perception. NetLogo is also a great tool for experimenting

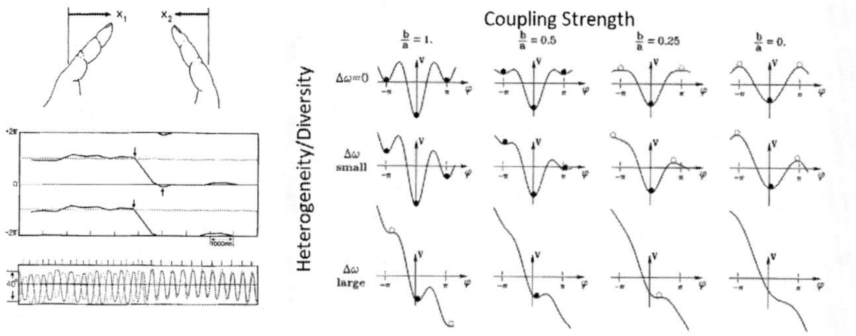

Figure 5.7. The finger-movement task. Two fingers move up and down (x_1 and x_2). They can move in phase or out of phase with a phase difference of 0 and π (bottom left figures). The model is shown on the right side. The potential function either has two stable states (a phase difference φ of 0 or π; $-\pi$ is the same state) or only one stable state (a phase difference of 0). Coupling strength, b/a, and heterogeneity, Δw, are control variables. Refer to the online version of the book for a more detailed image. *(Adapted from Haken, Kelso, and Bunz 1985; Kelso 2021.)*

with these effects. Download "Motion Quartet" from the NetLogo community website (or from this book's software repository) and explore hysteresis in your own perception.

5.3.4 MOTOR ACTION

Many body motions are periodic in nature—think of walking, swimming, dancing, and galloping. A famous paradigm for studying coordinative movement patterns is the finger movement task, in which one has to move both index fingers up and down (or right and left), either in phase or out of phase. Figure 5.7 explains the setup and data showing the transition between two in-phase or out-of-phase oscillations.

Key to these complex motions is the synchronization of the movements of body parts.

The Haken–Kelso–Bunz (HKB) model, developed in the tradition of synergetics, explains the phase transition between in-phase and antiphase motions in a way we saw in section 3.4. They set up a potential function in the form of

$$V(\varphi) = -\Delta w\varphi - b\cos\varphi - a\cos 2\varphi, \tag{5.4}$$

where φ is the order or behavioral variable, the phase difference between the two fingers. The main control parameter is b/a. According to Kelso (2021), coupling strength (b/a) corresponds to the velocity or frequency of the oscillations in the experiments. Δw is the difference (heterogeneity, diversity) between the natural frequencies of the individual oscillatory elements. In the finger-movement task, this parameter is expected to be 0. The behavior of this potential function is cusp-like. It has two stable states, 0 and $\pm\pi$, and increasing and decreasing the frequency leads to hysteresis. The effect of Δw is similar to the fold catastrophe (sec. 3.3).

This potential function is proposed as the simplest form that explains the experimental results. This is why I would call this a phenomenological model. However, Haken, Kelso, and Bunz (1985) also present a more mechanistic model, a combination of van der Pol and Rayleigh oscillators (Alderisio, Bardy, and di Bernardo 2016). The stochastic variant of the HKB model also features early warnings such as critical slowing down (see the catastrophe flags, sec. 3.5). The presence of critical slowing down and other flags has been confirmed experimentally (Kelso, Scholz, and Schöner 1986).

One difference with the catastrophe approach is that the synergetic models that incorporate hysteresis typically do not have a splitting control variable. The concept of structural stability, which is fundamental to catastrophe theory, is not used in synergetics. What the splitting factor might be in this model is not so clear. I have never understood why coupling strength b/a (see fig. 5.7) and the frequency of the oscillations are equated in the basic version of the HKB model (see also Beek, Peper, and Daffertshofer 2002). Clearly, uncoupled oscillators would have a rather random phase difference. Strengthening the coupling would lead to a kind of pitchfork bifurcation.

This coupling and uncoupling is also a phenomenon in the visual coordination of rhythmic movements between people.

Schmidt, Carello, and Turvey (1990) used an experimental paradigm in which two people swing a leg up and down while sitting side by side. A metronome was used to manipulate the frequency of the swing. Clear jumps from out-of-phase to in-phase movement were demonstrated.

Kelso (2021) provide an overview of the impressive amount of work on the HKB model. Repp and Su (2013) review empirical work in many different motor domains. Interestingly, learning motor tasks sometimes involves learning to couple movements (walking) and sometimes to uncouple movements (to drum more complex rhythms). Juggling is a fascinating case that has been studied in great detail (Beek and Lewbel 1995). Another popular mathematical approach to synchronization phenomena is the Kuramoto model (Acebrón et al. 2005) with the synchronous flashing of fireflies as a basic example. The Kuramoto model shows how synchronization depends on the coupling strength: below a certain threshold, the oscillators behave independently, while above this threshold, a significant fraction of the oscillators spontaneously lock to a common frequency, leading to collective synchronization. A second-order multi-adaptive neural agent model of interpersonal synchrony can be found in Hendrikse, Treur, and Koole (2023).

5.3.5 ROBOTICS

A major challenge in robotics is to build walking robots. Bipedal robots have evolved from clumsy mechanical walkers to flexible dynamic walkers and runners. Current legged robots can walk on uneven natural terrain, jump, do backflips, recover from rear shocks, and dance (see some videos on humanoid robots such as Atlas and Asimo). These successes are based on a combination of new technologies, but the principles of self-organization play a key role (Pavlus 2016). An important concept is dynamic stability. In old-school robots, the path and momentum of each step had to be precisely calculated in advance to keep the robot's center of mass continuously balanced at every point. Modern robots use sensory feedback systems to balance and adjust their movements on the fly, making them more adaptable to different and changing environments.

A dynamically stable robot maintains balance the same way a human does: by catching itself midfall with each step.

An intriguing application is called passive dynamics, which refers to robotic walking without external energy supply (McGeer 1990; Reher and Ames 2021). The idea is that truly dynamic locomotion should be based on the nonlinear dynamics in natural walking systems. An amazing demonstration is the artwork *Strandbeest* by Theo Jansen (fig. 5.8). Inspired by another great book about self-organization, *The Blind Watchmaker* (Dawkins 1986), Jansen created generations of kinetic sculptures made of PVC piping, wood, fabric wings, and zip ties that can move across the sand, resembling walking animals. His YouTube videos are recommended.

5.3.6 DEVELOPMENTAL PROCESSES

The early roots of interest in nonlinear dynamics and self-organization can be found in the groundbreaking work of French psychologist Jean Piaget. In order to understand the origin of knowledge, he studied the origin of intelligence in the child (Piaget 1952). His theorizing was inspired by both biological models and observations of children solving puzzles. He saw cognitive development as the building of structures on

Figure 5.8. Beach Beast © Theo Jansen, Umerus 2009, c/o Pictoright Amsterdam 2024.

earlier knowledge structures in a process of equilibration. The idea was that the child would assimilate or accommodate to potentially conflicting external information. In the case of assimilation, the child modifies the information to fit the current cognitive structure, while in the case of accommodation, the structure is modified. Such a modification could be the addition of an exception to the rule ("Longer sausages of clay normally weigh more, but not when this professor rolls the clay ball into a sausage," see sec. 3.1). In the long run, this does not work, the cognitive conflicts intensify, and the cognitive structure is destabilized. In this state of disequilibrium, a new structure can be formed on top of the earlier structure. An example of this is the conservation task I introduced in the introduction of chapter 3. The pre-operational structure, in which form and quantity are equated, leads to incorrect predictions in the conservation anticipation task. The child may ignore this (assimilation) and create an ad hoc rule for this exception (accommodation), but such solutions do not really resolve the cognitive conflict, and the pre-operational structure becomes unstable. This instability allows the formation of the more advanced concrete operational structure in which form and quantity are independent constructs.

Cognitive conflicts lead to a state of disequilibrium, resulting in the formation of new structures on top of the previous cognitive structure.

Piaget argued that cognitive development is a spontaneous, natural process that occurs as children interact with the world around them. I see my own work in developmental psychology (e.g., Savi et al. 2019; van der Maas et al. 2006; van der Maas and Molenaar 1992) as a formalization of these classical ideas of Piaget. The idea of stages and equilibrium lives on in neo-Piagetian theories.

Piaget's concept of cognitive development can be viewed as self-organization theory avant la lettre, as was the case with the Gestalt psychologists.

In the late twentieth century, developmental theories inspired by work in embodied cognition, nonlinear dynamics, synergetics, and neural networks (e.g., Edelman's neural Darwinism) became popular. Embodied cognition is the theory that an individual's understanding and thinking are intricately connected to the body's interactions with the environment, suggesting that cognitive processes are shaped by the body's actions and sensory experiences (Chemero 2013). A key example is Esther Thelen's work on the development of walking and reaching (Thelen 1995). Another famous Piagetian task, the A-not-B error, plays a central role in this. The A-not-B error typically occurs in a simple game where an adult hides an object in a known location (A) in front of an infant several times. After a few trials, the adult hides the object in a new location (B) while the infant is watching. Despite watching the object being hidden in the new location, infants tend to continue searching for the object in the old location (A).

Thelen and Smith's book (1994) had a strong influence on developmental psychology, although I was rather critical in my youthful enthusiasm (van der Maas 1995). Concrete mathematical dynamical models for A-not-B error have been developed in dynamic field theory (Schöner and Spencer 2016). These dynamic fields can be thought of as distributed representations that encode information about specific aspects of a task or behavior. For example, there may be a dynamic field representing the position of an object in space or the intended movement trajectory of a limb. In this theory, complex behaviors arise from the coordination and integration of multiple dynamic fields. Dynamic field theory is an active area of research.[6]

Dynamic field theory posits that cognitive processes are represented as dynamic fields, which are patterns of neural activity that evolve over time.

Finally, I note that some recent work considers the educational system itself as a complex system (Jacobson, Levin, and Kapur 2019; Lemke and Sabelli 2008).

[6] See https://dynamicfieldtheory.org

5.3.7 PSYCHOLOGICAL DISORDERS

Somewhat dated but interesting reviews of the application of the self-organization concept in clinical psychology are provided by Barton (1994) and Ayers (1997). Barton's review begins: "There is perhaps no other area in which chaos theory, nonlinear dynamics, and self-organizing systems are so intuitively appealing yet so analytically difficult as in clinical psychology." Ayers also concludes that most applications in this field have been rather metaphorical.

In recent work, both the modeling and the empirical work have become more concrete (G. Schiepek and Perlitz 2009). An example is the mathematical model of marriage (Gottman et al. 2002) discussed in section 4.3. Tschacher and Haken (2019) present a new approach to psychotherapy based on complex-systems theory. They integrate deterministic and stochastic forces using a Fokker–Planck mathematical approach.

This network approach to psychological disorders suggests that psychological disorders arise from complex interactions among symptoms, rather than being caused by a single underlying factor.

In section 6.3 I introduce the network approach to psychopathology (Borsboom 2017; Cramer et al. 2010). It views disorders as interconnected networks of symptoms, where each symptom influences and is influenced by other symptoms. This approach emphasizes the dynamic nature of psychological disorders and highlights the importance of understanding the relationships between symptoms in order to effectively diagnose and treat them. Network modeling is accompanied by a new family of statistical techniques (Epskamp, Borsboom, and Fried 2018). An introduction to these techniques is given in section 6.4.

Recent reviews of the complex-systems approach to psychological and psychiatric disorders are provided by Olthof et al. (2023) and Scheffer et al. (2024).

5.3.8 SOCIAL RELATIONS

A key publication in this area is *Dynamical Systems in Social Psychology*, edited by Vallacher and Nowak (1994). Concepts such as dissonance (Festinger 1962), balance (Heider 1946), and harmony (Smolensky 1986) reflect the idea that we optimize internal consistency when forming attitudes and knowledge. A formal implementation of these ideas was proposed using parallel distributed processing–type connectionist models (e.g., Monroe and Read 2008). Our own model (Dalege and van der Maas 2020; Dalege et al. 2018, 2016) is based on the Ising model and the Boltzmann machine, as in Smolensky's proposal, which can be fitted to data. I will explain this work in more detail in the next chapter (sec. 6.3).

A famous example of social self-organization concerns pedestrian dynamics as studied by Helbing and Molnár (1995). They proposed a physics-based model for panic evacuation. For an excellent overview of crowd simulation, I again refer to Wikipedia. Some of this work is rooted in the social sciences. An example in NetLogo is the model "Path."

Also famous is the work of the sociologist Mark Granovetter (1973) on strong and weak ties in social networks (belonging to the most-cited papers in the history of the social sciences). Weak ties provide access to new information and opportunities that may not be available within one's close circle of friends and acquaintances. He also contributed the threshold model for collective action (Granovetter 1978). I like to explain this work using the "Guy starts dance party" video on YouTube. The idea is that people have some threshold, between 0 and 1, to join the dancers. The thresholds are sampled from the beta distribution, which is a flexible distribution determined by two shape parameters, α and β. With this R code we can simulate this effect:

Weak ties in social networks are often more valuable than strong ties.

```
layout(1:2)
n <- 1000 # number of persons
iterations <- 50
# sample individual thresholds for dancing:
threshold <- rbeta(n, 1, 2)
hist(threshold, col = 'grey')
dancers <- rep(0, n) # nobody dances
dancers[1] <- 1 # but one guy
number_of_dancers <- rep(0, iterations)
for(i in 1:iterations){
  # keep track of number of dancers:
  number_of_dancers[i] <- sum(dancers)
  # if my threshold < proportion of dancers, I dance:
  dancers[threshold < (number_of_dancers[i]/n)] <- 1
}
plot(number_of_dancers, xlab = 'time', ylab = '#dancers',
  ylim = c(0, 1000), type = 'b', bty = 'n')
```

Depending on the parameters of the beta distribution, you will see a phase transition to collective dancing. This basic setup can be extended in many ways.

Another classic contribution, explained in more detail in section 7.2, is Schelling's agent-based model of segregation (Schelling 1971). The idea is that even if individuals have only a small preference for in-group neighbors, segregated societies will form. For a broad overview of complex-systems research on human cooperation, I refer to Perc et al. (2017). A recent book on modeling social behavior using NetLogo is written by Smaldino (2023).

5.3.9 COLLECTIVE INTELLIGENCE

The wisdom of crowds posits that the collective judgments of a large group of people can be more accurate and effective than those of a single expert or small group.

Collective-intelligence research examines how groups can collectively outperform individual members in problem-solving, decision-making, and idea generation. One famous concept is the idea of the wisdom of crowds (Surowiecki 2005). A key example is the "Guess the Weight of the Ox" contest that took place at the West of England Fat Stock and Poultry Exhibition in 1906. While individual guesses varied widely, the median guess was remarkably close to the actual weight of the ox. The average guess was only one pound off the actual weight, which was 1,198 pounds (Galton 1907).

However, there is a fine line between the wisdom of the crowd and the stupidity of the crowd. It is extremely useful to know when that line is crossed. The wisdom of crowds tends to work when there is a diverse group of independent individuals, each making their own judgments or estimates about a particular question or problem (Brush, Krakauer, and Flack 2018; Centola 2022). Path dependency on previously faced problems and solutions might also play a role (Galesic et al. 2023). There is an extensive and up-to-date Wikipedia on collective intelligence, discussing findings from various disciplines, biological examples (swarm intelligence), and an overview of applications (such as open-source software, crowd sourcing, the Delphi technique, and Wikipedia itself).

Collective intelligence is more likely to be effective when the group is large, has a wide range of knowledge and perspectives, and makes judgments independently.

5.3.10 GAME THEORY

Game theory consists of mathematical models of strategic interactions among rational agents. A great historical overview can be found at Wikipedia. One of the most famous paradigms is the prisoner's dilemma. You and your friend are arrested, and you both independently talk to the police. The options are to remain silent or to talk. The dilemma is that remaining silent is the best option if you both choose it, but the

B / A	B stays silent	B betrays
A stays silent	-2 / -2	0 / -10
A betrays	-10 / 0	-5 / -5

Figure 5.9. Standard prisoner's dilemma payoff matrix. If both A and B remain silent, they each face a two-year sentence. If one talks and the other does not, the informer is released and the silent partner gets a decade behind bars. If both betray, they serve five years.

worst option if your friend betrays you (see the payoff matrix, fig. 5.9). In this game, loyalty to one's friend is irrational, an outcome related to the tragedy of the commons (Hardin 1968). The tragedy of the commons can be studied in the hubnet extension of NetLogo, where multiple users can participate in NetLogo simulations.

The tragedy of the commons occurs when individuals, acting in their own self-interest, overexploit a shared resource, leading to a depletion that undermines everyone's long-term interests, including their own.

A major topic in game theory is altruism. In many cases, individualistic choices lead to an unsatisfactory Nash equilibrium. The public-goods game is a good example. In this game, everyone invests some money, which is then multiplied by an external party (the government). Then everyone gets an equal share of the multiplied total. The problem is that free riders, who do not invest, win the most, which in iterated public-goods games leads to a situation where no one invests and no one wins. Punishment (shaming and blaming) is known to help combat free riding. But punishment also requires investment. I like to tell my students, when they are working in groups on an assignment, that the problem of this one student doing nothing happens because nice, hardworking students refuse to betray their fellow students. These nice, hardworking students are what are called second-order free riders (Fowler 2005). Just so you know.

A Nash equilibrium is a set of strategies in which no player can improve their payoff by unilaterally changing their strategy, given the strategies of the other players.

5.3.11 SELF-ORGANIZATION IN ORGANIZATIONS

Translating this basic research into real-world applications is far from straightforward (Anderson 1999b; Morel and Ramanujam 1999). Our economic system is a mixture of self-organization (pure capitalism) and

Human organizations can be placed on a scale from extreme hierarchy to radical forms of self-organization.

top-down regulation (through laws, taxes, and other regulations) (Volberda and Lewin 2003). Black markets are critical cases of unregulated self-organized systems (Tesfatsion 2002).

A concrete modeling example is the team assembly model by Guimerà et al. (2005). They study how the way creative teams self-assemble determines the structure of collaboration networks. The idea is that effective teams find a balance between being large enough to allow for specialization and efficient division of labor among members, and small enough to avoid excessive costs associated with coordinating group efforts. Agents in the model have only a few basic characteristics that influence their behavior: whether they are a newcomer or incumbent and what previous connections they have with other agents if they are incumbents.

Three parameters can be adjusted to influence behavior in the baseline assembly model: the team size, the probability of choosing an incumbent (p), and the probability of choosing a previous collaborator (q). The two probability parameters signify assumptions about agent motivations for team member selection. Low incumbent probability leads to preference for newcomers and new ideas, while high incumbent probability means a focus on experience. Low collaborator probability prioritizes experienced strangers, and high collaborator probability prioritizes previous collaborators. The model is part of the built-in NetLogo Model Library ("Team Assembly"). By simulating the model, it can be shown that the emergence of a large, connected community of practitioners can be described as a phase transition (fig. 5.10).

Guimerà et al. (2005) estimated the parameters p and q for the community formation in four scientific disciplines (social psychology, economics, ecology, and astronomy). Only astronomy had a very dense collaboration structure. In the other fields, the estimates of p and q of teams publishing in certain journals correlated well with impact factor. Interestingly, p correlates positively and q negatively with impact.

5.5 Zooming Out

I hope I have succeeded in giving an organized and practical overview of a very disorganized and interdisciplinary field of research. For each subfield, I have provided key references that should help you find recent

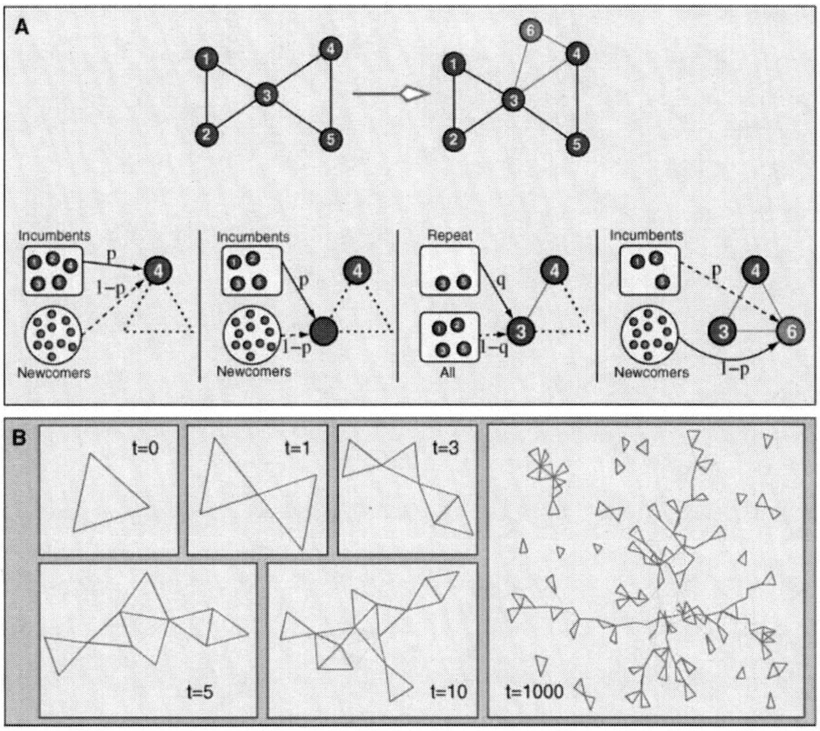

Figure 5.10. Team assembly model. Newcomers and incumbents are added to growing networks based on probabilities p and q. If p is sufficiently high, a dense network emerges. Refer to the online version of the book for a color image. *(Adapted from Guimerà et al. (2005) with permission.)*

and specialized contributions. I find the examples of self-organization in the natural sciences fascinating and inspiring. I hope I have also shown that applications of this concept in psychology and the social sciences hold great promise. In the next chapters, I will present more detailed examples.

I believe that understanding models requires working with models, for example, through simulation. NetLogo is a great tool for this, although there are many alternatives (Abar et al. 2017). I haven't mentioned all the uses of NetLogo, but it's good to know about the BehaviorSpace option. BehaviorSpace runs models repeatedly and in parallel (without visualization), systematically varying model settings and parameters, and recording the results of each model run. These results can then be further analyzed in R. An example is provided in chapter 7, section 7.2.

I have largely omitted the network approach in this chapter. Psychological network models are a recent application of self-organization in complex systems in psychology and are the subject of the next chapter.

5.6 Exercises

1. Is there a relation between the rice cooker and the Ising model? How does the magnetic thermostat in a traditional rice cooker work to automatically stop cooking when the rice is done? (*)

2. What is the Boltzmann entropy for the state $\sum x = 0$ in an Ising model (with nodes states -1 and 1) with 10 nodes and no external field? (*).

3. Go to the web page "A Neural Network Playground" (https://playground.tensorflow.org). What is the minimal network to solve the XOR close to perfect accuracy? Use only the x1 and x2 feature. (*)

4. In the Granovetter model (sec. 5.4), people may also stop dancing (with probability .1). Add this to the model. How does this change the equilibrium behavior? (*)

5. Add the external field to the Ising model in NetLogo (neighbors4 case). Report the changed line in the NetLogo code. What did you change in the interface?
   Set the temperature to 1.5. Change `tau` slowly. At which values of `tau` do the hysteresis jumps occur? (*)

6. Test whether the Ising model is indeed a cusp. Run the Ising model in NetLogo using the BehaviorSpace tool (see fig. 7.1 for an example). Use the model in which all spins are connected to all spins (see sec. 5.3). Vary `tau` ($-.3$ to $.3$ in $.05$ increments) and `temperature` (0 to 3, in $.5$ increments). One iteration per combination of parameter values is sufficient. Stop after $10,000$ ticks and collect only the final magnetization. Import the data into R and fit the cusp. Which cusp model best describes the data? (**)

7. Open the Sandpile 3D model in NetLogo3D. Grains of sand fall at random places. Change one line of code such that they all fall in the middle. What did you change? (*)

8. Download "Motion Quartet" from the NetLogo community website and explore hysteresis in your own perception. What could be a splitting variable? (*)

9. Implement the Granovetter model in NetLogo (max 40 lines of code). (**)

10. Implement Game of Life in NetLogo or use Golly and try to find as many qualitatively different stable patterns of six units that can occur in Game of Life. If you cannot find more, try to look at additional resources online to find the other patterns you missed. For four units, there are only two, one of which is a block of four. (*)

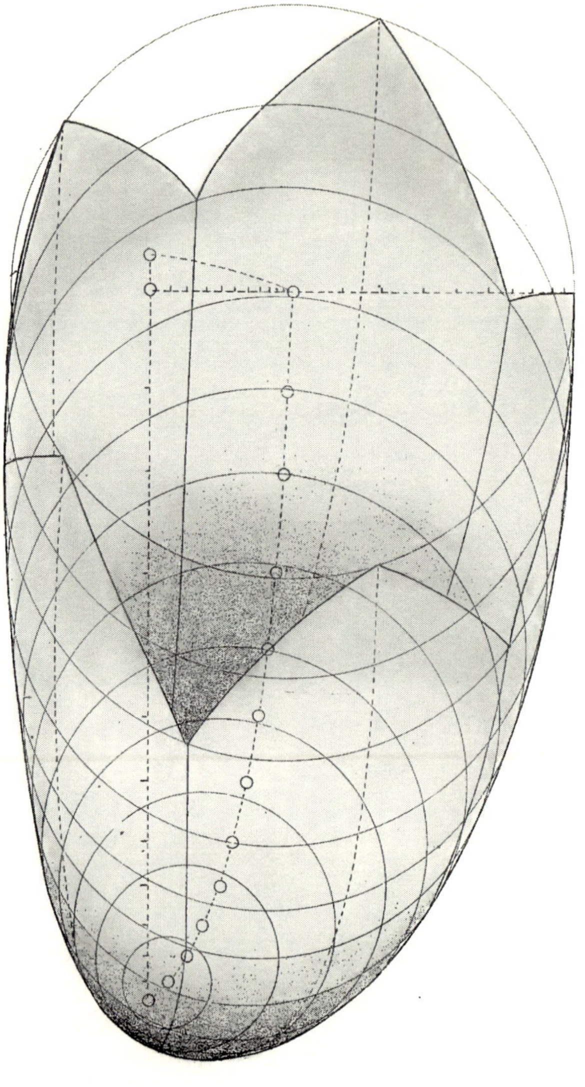

8

PSYCHOLOGICAL NETWORK MODELS

6.1 Introduction

Common-cause theories are widely accepted in psychology. High scores on cognitive tests are attributed to high intelligence. Charisma is associated with various leadership qualities. A high score on the p factor (psychopathology factor) is associated with various mental health problems. The common-cause explanation is popular because of its simplicity and the availability of statistical methods such as factor analysis. In most cases, however, the common cause is only a hypothetical construct. Yet many people accept common-cause theories because they see no alternative. This is incorrect. Complex-systems theory offers a powerful alternative, including a statistical approach. The main thesis of this chapter is that the functioning (and dysfunctioning) of the human mind is often best understood as a complex interplay of various psychological elements such as cognitive functions, mental states, symptoms, and behaviors.

RESOURCES
The electronic version of this book, including code blocks and links, can be viewed and accessed at: *santafeinstitute.github.io/ComplexPsych/*

Common-causes are questionable if they cannot be identified independently of the observed relationships they are intended to explain.

The interplay of psychological subsystems can be modeled in terms of networks: psychological networks.

The study of psychological networks is probably the most thriving area of complex-systems research in psychology today. This chapter is about this new line of research. The paper on the mutualism model of general intelligence (van der Maas et al. 2006) can be seen as the root of this approach, but it took off, as shown in figure 6.1, when it was applied to clinical psychology (Borsboom 2017, 2008; Cramer et al. 2010), especially when the theoretical work was backed up with psychometric tools (Epskamp, Borsboom, and Fried 2018; Epskamp et al. 2012; Marsman and Rhemtulla 2022).

In this chapter I will present and discuss these theoretical and psychometric lines of research, accompanied by practical examples. First, I will begin with an introduction to network theory.

6.2 Network Theory

It is hard to imagine a discipline in which networks are not a central theme. Networks are the key to understanding systems ranging from

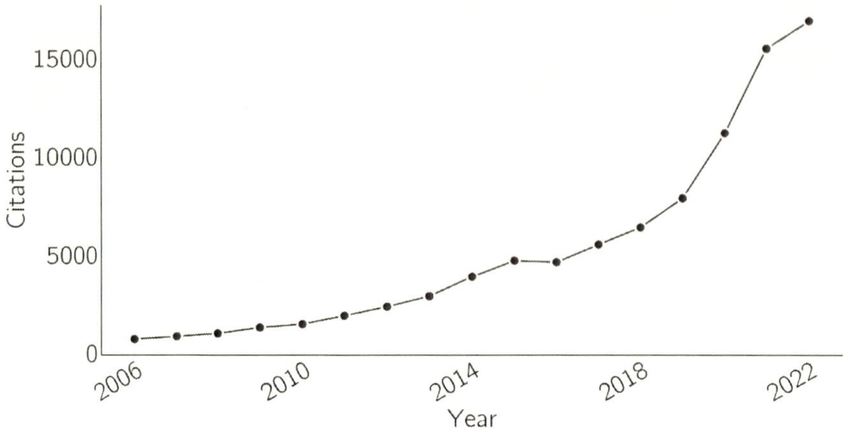

Figure 6.1. The number of papers in Google Scholar on the combination of symptoms and "network analysis" grows exponentially.

particle physics to social networks, from ecosystems to the internet, and from railways to the brain. The mathematics of network theory is not so easy to grasp, but fortunately the basic concepts are.

6.2.1 NETWORK CONCEPTS

A network, a special type of graph, consists of nodes (often called vertices) and links (or connections or edges).

Nodes can be anything—particles, neurons, words, people, train stations, etc. The size of a network is equal to the number of nodes. Nodes are connected by links. Links can be directed or undirected. For example, causal links are directed. Occasionally you see links from the node to itself. In causal networks, this may represent self-excitation, or if the weight is negative, self-inhibition. In some cases, links are simply present or absent; in other cases, links are weighted, as in most neural networks. The matrix of all weights, indicating the strengths of the connections between nodes, is called the adjacency or edge matrix. For an undirected network, the adjacency matrix is symmetric. For a network without self-loops, the diagonal of the adjacency matrix is 0.

A connected network is a network in which every node is connected to every other node, possibly through intermediate nodes. In a fully connected network, or complete graph, every node is directly connected to every other node. Such a network has a density of 1 (i.e., the proportion of edges that is present).

Nodes can be in the center of a network or in the periphery. This should not be taken literally as psychological networks have no spatial dimension. There are many kinds of centrality measures, such as closeness centrality and degree centrality. The degree of a node is equivalent to the number of links it has. The average degree of a network is the average of the number of links over all nodes.

Centrality measures quantify the relative influence, control, or connectivity of a node compared to other nodes in the network.

The degree distribution can take several forms. A random graph, where nodes are connected randomly, has a binomial degree distribution. Most real-world networks have a skewed degree distribution. The average shortest path length (ASPL) is the average number of edges that must be traversed to get from one node to another using the shortest paths (i.e., the fewest intermediate nodes).

There are many methods available in R for creating and visualizing networks and for computing properties of networks. The igraph and qgraph libraries are very useful. It is a good idea to experiment with this R code below by varying the parameter values. Some key concepts of networks are shown in figure 6.2.

```
library(igraph); library(qgraph)
g1 <- graph(edges = c(1,2, 2,3, 3,1),
            n = 3, directed = FALSE)
plot(g1) # an undirected network with 3 nodes
g2 <- graph(edges = c(1,2, 2,3, 3,1, 1,3, 3,3),
            n = 3, directed = TRUE)
plot(g2) # a directed network with self-excitation on node 3
get.adjacency(g2) # weight matrix
fcn <- make_full_graph(10) # a fully connected network
plot(fcn, vertex.size = 10, vertex.label = NA)
layout(1); set.seed(1)
# a weighted adjacency matrix:
adj <- matrix(rnorm(100, 0, .2), 10, 10)
# set 80% to 0:
adj <- adj * sample(0:1, 100, replace = TRUE, prob = c(.8, .2))
qgraph(adj) # plot in qgraph
edge_density(fcn) # indeed 1
edge_density(graph_from_adjacency_matrix(adj, weighted = TRUE))
centralityPlot(qgraph(adj))
# note centrality() gives more indices
```

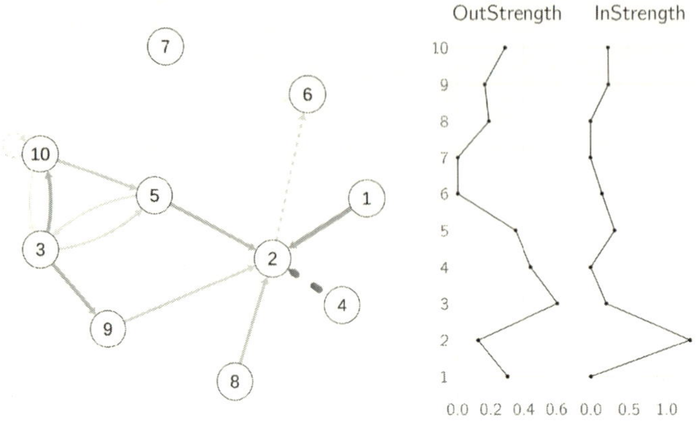

Figure 6.2. A weighted directed network with self-loops. Red arrows indicate negative effects. OutStrength and InStrength represent two types of centrality measures in directed networks. Refer to the online version of the book for a more detailed color image.

6.2.2 NETWORK TYPES

Simple networks do not have cycles. An example of an undirected acyclic graph is one with nodes on a line (but not a circle). Connected undirected acyclic graphs are trees; if they are partially unconnected, they are forests. Directed acyclic graphs, such as family trees and citation networks, are called DAGs. Most current neural networks are feedforward networks without cycles, but so-called recurrent networks have cycles.

Igraph[1] has an amazing number of functions for creating specific networks. Some examples are shown in figure 6.3.

The last case in the figure, preferential attachment, is of particular interest because it is created dynamically. It starts with a node and then new nodes are added that prefer links to nodes that already have many links. In the NetLogo model "Preferential Attachment simple," you can see this growth process. Preferential attachment networks are often called complex because they exhibit nontrivial structural patterns. Preferential attachment networks are "scale-free," meaning that the degree distribution looks the same no matter the scale. Scale-free networks are useful for studying the robustness and vulnerability of networks to targeted attacks on highly connected nodes. Removing hubs with high connectivity potentially split the network into disconnected components and impede the network's func-

Scale-free networks have a degree distribution that follows a power law, with some nodes having many links but most having only a few.

[1] https://igraph.org

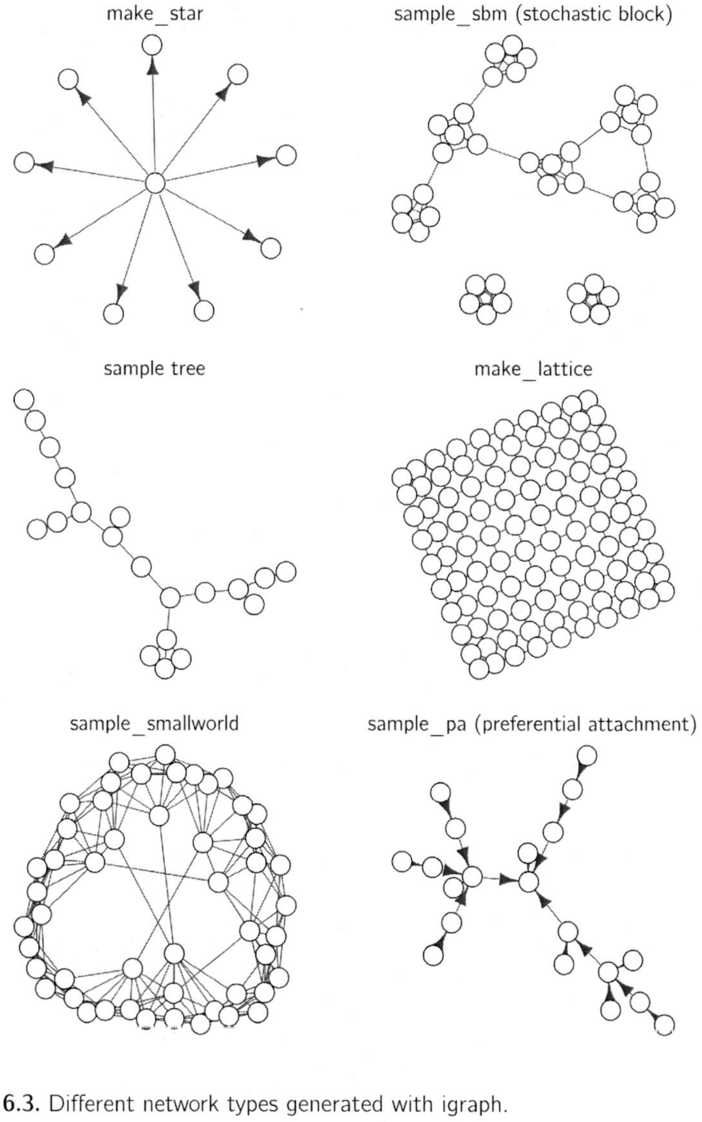

Figure 6.3. Different network types generated with igraph.

tionality. Because most nodes in the network have a small degree (few connections), randomly removing nodes tends not to disrupt the network's overall structure. Scale-free networks are believed to exist in various real-world scenarios, ranging from website connections to scientific collaborations. For a critical analysis, I refer to Broido and Clauset (2019).

Small-world networks consist of clusters, but there are also links between the clusters.

Another type of complex network is the small-world network. The distance between any two nodes in such a network is always relatively short. A famous example is the six-handshake rule (also known as the six degrees of separation), which states that all people are six or fewer handshakes away from each other. [2] For this reason, small-world networks are useful for studying the spread of information or disease through social networks.

The scale-free and small-world networks are predominantly associated with the complex-systems approach. However, I believe that the hierarchical or nested stochastic block model (HSBM) is equally relevant (Clauset, Moore, and Newman 2008). The HSBM extends the SBM concept: clusters are nested within larger clusters, which in turn are part of even larger clusters in a continuous sequence (see fig. 6.4), resembling fractals.

In the stochastic block model (SBM), nodes are organized into clusters with connections being stronger or more frequent within these clusters than between them.

This nesting seems to be crucial for understanding complex systems and is a central theme in Herbert Simon's influential architecture of complexity (Simon 1962). He introduced the concept of near decomposability to describe the interaction within these nested hierarchies. Typically, interactions within each subsystem are stronger and more frequent than those between subsystems. Although the HSBM simplifies reality, where levels can intermingle and low-level interference might occasionally escalate to higher levels, it often serves as a useful framework for conceptualizing complex networks, including in the field of psychology.

6.2.3 NETWORK DYNAMICS

In a random graph or network, as the density of edges increases beyond a certain threshold, a phase transition occurs where a giant component suddenly appears.

A prominent phase transition in network theory is the emergence of a giant component. This happens when we start with a completely unconnected network of n nodes and randomly add links. We simply take a random node and connect it to another node to which it has no connection. This leads

[2] Amazingly, most people are only four or five handshakes away from Napoleon.

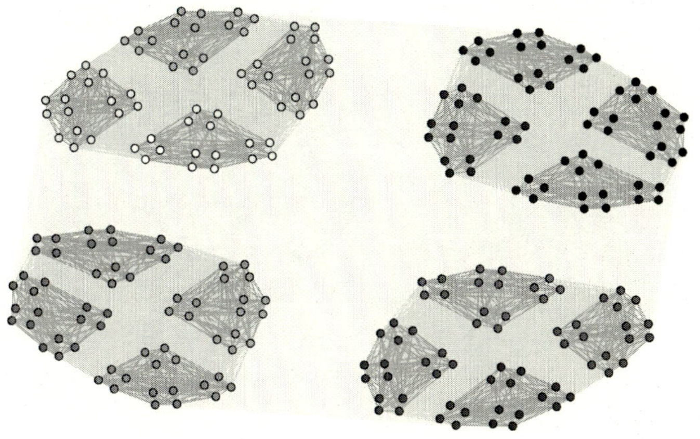

Figure 6.4. A hierarchical stochastic block model with four levels of four units. Probabilities of links are higher within blocks than between blocks of nodes. This embedding of levels could be seen as an implementation of Simon's architecture of complexity. The code for this figure is available in the online software repository of the book.

to many small unconnected clusters at first, but then a giant component appears (a second-order phase transition). This happens when about $n/2$ links have been added. You can verify this in the NetLogo model "Giant Component." The implication is that randomly connected networks with a sufficient number of links are almost always connected networks.

This is just one example of network dynamics (Dorogovtsev and Mendes 2002). We can distinguish between dynamics on node values (e.g., Lotka–Volterra models), on link values (connection strength in neural networks), and cases where the structure of the network is dynamic, as in the giant component example. These types of dynamics also coexist and interact. In neural networks, both node and link values are updated (on fast and slow time scales). We will see more examples in the next chapter.

A relatively new topic in complex networks concerns higher-order interactions. In most networks, we only consider pairwise interactions, but third-order and even higher-order interactions may play a role (Battiston et al. 2021). Other work considers hierarchical complex networks (Boccaletti, Bianconi, Criado, Genio, et al. 2014). For more information on network concept and types, I first refer to Wikipedia. Another great (open) source is the book by Barabási and Pósfai (2016). A more concise overview is provided by Boccaletti, Bianconi, Criado, Genio, et al. (2014).

6.3 Psychological Network Models

Differential psychology is concerned with individual differences, in contrast to experimental psychology, which is concerned with mechanisms. This division comes from a renowned paper by Cronbach (1957) on the two disciplines of scientific psychology. Cronbach distinguished between the how question (how does one read a sentence) and the why question (why do we differ in reading). The latter is typical of differential psychology. The latent variable or factor approach has long been dominant in differential psychology. When studying individual differences in a trait, psychologists generally follow the same approach. They construct tests, collect data, perform factor analysis, and propose one or more latent traits to explain observed individual differences. The justification for this approach, particularly in intelligence research, rests primarily on its predictive power (van der Maas, Kan, and Borsboom 2014).

Latent variables are used in statistical modeling to represent unobservable or underlying factors that cannot be directly measured or observed.

The statistical tools for analyzing latent variables come from modern test theory and structural equation modeling (SEM). These technically advanced tools are developed in a field called psychometrics. However, despite this technical sophistication, it is often not clear what latent variables are in psychometric models. Some researchers tend to think of them as purely statistical constructs that help summarize relationships between variables and make predictions. But more often, either implicitly or explicitly, latent variables are interpreted as real constructs, as common causes of observed measures (van Bork et al. 2017). The psychological network approach was developed in response to the factor approach. The main motivation for the network approach is that underlying common causes are unsatisfactory if they cannot be identified independently of the observed relationships they are supposed to explain (van der Maas et al. 2006). One consequence is that such an explanation does not provide guidance for possible interventions.

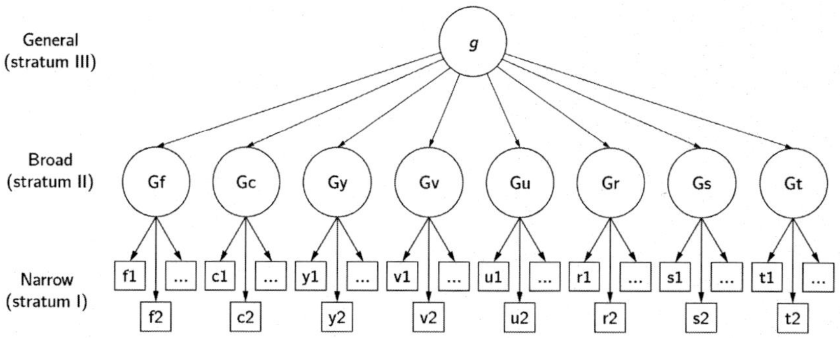

Figure 6.5. The Cattell–Horn–Carroll model of general intelligence. Blocks represent test scores (narrow), explained by the broad factors, which in turn are determined by the general factor g.

6.3.1 MUTUALISM MODEL: THE CASE OF GENERAL INTELLIGENCE

6.3.1.1 The g Factor

The factor-analysis tradition in psychology began with the study of general intelligence, and so does the psychological network approach. The factor or g model of general intelligence was proposed by Spearman (1904) as an explanation of the positive manifold, that is, the much-replicated effect that subtests of intelligence test batteries are positively correlated. In the original simplest model, the observed test scores are statistically explained by a common factor, basically meaning that the correlations between test scores disappear when subjects have the same score on the common factor. In the Cattell–Horn–Carroll (CHC) model, often referred to as the standard model, test scores load on subfactors such as visual processing (Gv) and fluid reasoning (Gf), which in turn are positively correlated. These latent correlations are explained by the general, higher-order factor g (fig. 6.5).

This model has been criticized extensively, including for its alleged implications for group differences in observed IQ and intervention strategies (Fraser 2008). In my view, some of the criticism is unwarranted. For example, the positive manifold is a very robust and widely replicated empirical phenomenon (Nisbett et al. 2012). The specific tests included are not of great importance. That is, any reliable measure of creativity, emotional intelligence, or social intelligence correlates positively with other

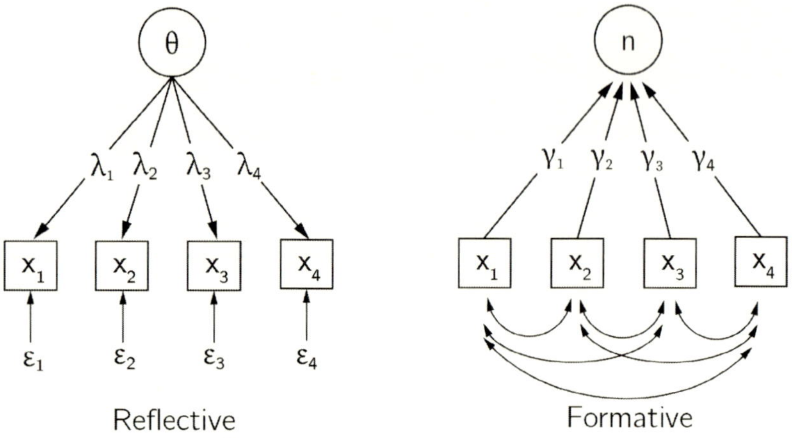

Figure 6.6. The reflective and formative interpretations of the factor model cannot be distinguished with correlational data, but they are very different. In the reflective model, the latent factor is a common cause (e.g., temperature) that causes the observations (e.g., different thermometers). Intervening on one thermometer (heating) will only change that particular thermometer because each x has no outgoing connections. In the formative interpretation, the factor is just an index (e.g., an economic index) that summarizes the state of many interacting components (companies). In this case, *only* interventions on the x can have an overall effect.

IQ subtests. Nor is there much wrong with factor analysis as a statistical technique. To me, the most questionable aspect of g theory is that it is not really a theory at all. The "elephant in the room" question is simply: What is g? What could this single factor be that explains everything? A century of research has not produced a generally accepted answer to this question. And this is a problem for many factor explanations in psychology (e.g., the big five of personality, the p factor of psychopathology).

It is important to note that the factor explanations are not problematic in and of themselves. I like to use the example of heart disease, say, a loose heart valve. This leads to symptoms such as shortness of breath, swelling of the ankles, dizziness, rapid weight gain, and chest discomfort. The relationship between these symptoms is explained by the underlying factor of heart disease. Treating a single symptom may provide some relief for that symptom, but not more. Only intervening on the cause will bring about real change. This is an example of a reflective interpretation of the factor model. When the factor is merely an index and not a common cause, we speak of a formative factor. Figure 6.6 explains the reflective and formative interpretations of the factor model.

Statistically, these factor models are equivalent. Thus, the fact that factor models fit intelligence data well does not tell us anything about the status of statistical factor g.

Is g a common cause or just an index?

6.3.1.2 Mutualism Model

In van der Maas et al. (2006), we proposed an alternative model that is consistent with the formative interpretation of the factor model. The idea is that our cognitive system consists of many functions that develop over time in an autocatalytic process based on experience and training but also due to weak positive reciprocal interactions between developing cognitive functions (fig. 6.7). Examples of such mutualistic interactions are those between short-term memory and cognitive strategies, language and cognition (syntactic and semantic bootstrapping), cognition and metacognition, action and perception, and performance and motivation (van der Maas et al. 2017). For example, babies learn to grasp objects by repeatedly reaching out, coordinating their hand and finger movements, and adjusting their grip. Through these actions, they gather sensory feedback, refining their perception and improving their grasping skills in a reciprocal learning process (Needham and Nelson 2023).

These mutualistic interactions can create correlations that are typically associated with factor models.

To model this, we used the mutualistic Lotka–Volterra model:

$$\frac{dX_i}{dt} = a_i X_i \left(1 - \frac{X_i}{K_i}\right) + a_i \sum_{\substack{j=1 \\ j \neq i}}^{W} \frac{M_{ij} X_i X_j}{K_i} \quad for\ i = 1..W, \tag{6.1}$$

$$K_i = c_i G_i + (1 - c_i) E_i,$$

where $X_i \ldots X_w$ denote the cognitive processes, **a** the growth rates, **K** the limited resources for each X_i (a weighted sum of a genetic (**G**) and an environmental (**E**) part), and **M** the interaction matrix. The second equation, assuming simple linear effects of genetics and environment, is sufficient to explain some typical phenomena in twin research, such as the increase in heritability with age (see van der Maas et al. 2006). Criticisms and tests of the mutualism models, as well as alternatives, are discussed in van der Maas et al. (2017). Knyspel and Plomin (2024) compare the mutualism model

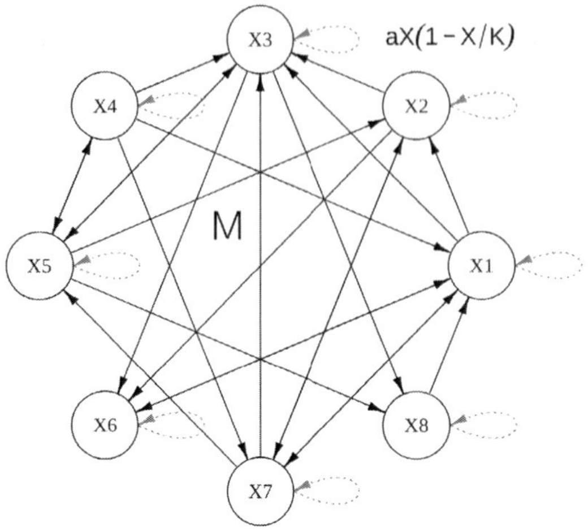

Figure 6.7. The mutualism model. The self-loops have an excitatory (aX) and an inhibitory part ($-aX^2/K$).

with the factor model using twin data. Here we focus on the technical aspects.

When M contains mostly negative values, the model is known as a competitive Lotka–Volterra model. In this case, limit cycles and other nonlinear phenomena may occur (Hirsch 1985). For the mutualistic variant, with positive M, we see either convergence to a positive state or exponential growth. This exponential growth is an unfortunate aspect of the Lotka–Volterra mutualism model. Robert May famously described this effect as an orgy of mutual benefaction (May, Oxford, and McLean 2007), which is not what we see in nature, and all sorts of solutions have been proposed (Bascompte and Jordano 2013).

The mutualism model in Grind is specified as follows:

```
mutualism <- function(t, state, parms){
  with(as.list(c(state, parms)),{
    X <- state[1:nr_var]
    # using matrix multiplication:
    dX <- a * X * (1-X/k) + a * (X * M %*% X)/k
    return(list(dX))
  })
}
```

A simulation of the positive manifold requires us to run this model for multiple people and collect the X-values after some time points (`tmax = 60`) for each person. We can then compute the correlations and check if they are positive (fig. 6.8). For each person, we resample a, K, and the initial values of X, but M is the same across persons. Note that the M-values should not be set too high, otherwise we end up in May's orgy of mutual benefaction. In the second part of this chapter, we will generate more data with this model and fit network and factor models.

```
layout(matrix(1:2, 1, 2))
nr_var <- 12 # number of tests, abilities (W)
nr_of_pp <- 500
# to collect the data in the simulation:
data <- matrix(0, nr_of_pp, nr_var)
M <- matrix(.05, nr_var, nr_var)
M[diag(nr_var) == 1] <- 0 # set diagonal of M to 0

for(i in 1:nr_of_pp){
  # sample a, K, starting values X from normal
  # distributions for each person separately
  # note M is constant over persons.
  a <- rnorm(nr_var, .2, .05)
  k <- rnorm(nr_var, 10, 2)
  x0 <- rnorm(nr_var, 2, 0.1) # initial state of X
  s  <- x0; p <- c() # required for grind
  # collect data (end points) and plot person 1 only
  data[i,] <- run(odes = mutualism ,tmax = 60,
      timeplot = (i == 1), legend = FALSE)
}
hist(cor(data)[cor(data) < 1], main = 'positive manifold',
     xlab = 'between test correlations',
     col = 'grey50') # positive manifold
```

6.3.1.3 Abnormal Development

In van der Maas et al. (2017), this model is applied in several ways, for example, by incorporating Cattell's idea of investment of fluid skills in crystalized abilities (discussed in sec. 4.5). In a recent paper, de Ron et al. (2023) extend the mutualism model with resource competition to explain different patterns of abnormal development. In the process of modeling, we came to an interesting insight. Assuming that there is competition for scarce resources (time, money, educational support), hyperspecialization might be

One pattern of abnormal development is hyperspecialization, which is associated with rare variants of autism.

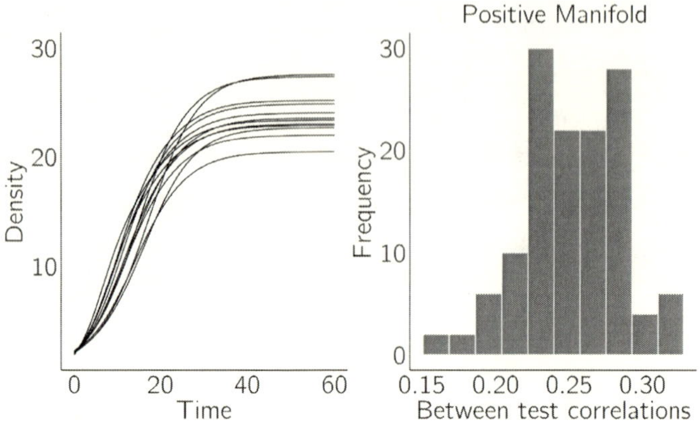

Figure 6.8. A typical run of the mutualism model for one subject and the distribution of correlations between X-values across subjects.

the default outcome, and thus it is "normal" development that needs to be explained. The reason is an insight from mathematical biology: ecosystem diversity is often unstable. An example we have already seen is hypercycle instability due to parasites (sec. 5.2). This is normally studied in resource competition models.

Ecosystem diversity is often unstable.

In basic resource competition models in population biology (Tilman, Kilham, and Kilham 1982), the growth of a species $(1...W)$ is determined by its current size X_i and the sum over resources $R_j(1...V)$. The parameters μ_{ij} determine how much species i benefits from the resource j. If no resources are available, X_i dies out with death rate d_i.

The growth of the resource R_j consists of two parts. The first part models the growth by a concave function, which is determined by r (i.e., the steepness of the concave function) up to $r_{\max}$. The second part is the depletion by consumption of resources by X_i at rates b_{ij}. Two differential equations specify these dynamics (see the appendix of de Ron et al. 2023 for the Grind code to study this model numerically):

$$\begin{aligned} \frac{dX_i}{dt} &= X_i\Big(\sum_{j=1}^{V} \mu_{ij} R_j - d_i\Big), \\ \frac{dR_j}{dt} &= r\,(r_{\max} - R_j) - R_j \sum_{i=1}^{W} b_{ij} X_i. \end{aligned} \tag{6.2}$$

What has been shown for this and related models is that you will not get more species surviving than there are resources. Another famous quote from Robert May is "There is no comfortable theorem assuring that increased diversity and complexity beget enhanced community stability; rather, as a mathematical generality, the opposite is true. The task, then, is to elucidate the devious strategies which make for stability in enduring natural systems. There will be no one simple answer to these questions" (p. 174, 2001 edition). Thus, given a limited number of resources (time, money, educational support), we should expect early specialization in only a few skills.

Biologists have proposed a number of mechanisms to deal with this problem (Meena et al. 2023). In de Ron et al. (2023), we added three mechanisms: density-dependent growth (see sec. 4.2) of the abilities X with a logistic term; mutualism between abilities as in the mutualism model; and growth-dependent depletion of resources. The idea of the latter is that the growth of abilities costs a lot of resources, but the maintenance much less. Learning arithmetic or chess requires a lot of effort, but once a certain level of mastery is reached, it remains roughly at that level without further training (unfortunately, this is not the case with physical condition).

We show that the combination of these mechanisms allows a balanced growth of several correlated abilities. Specially chosen parameter settings lead to different patterns of abnormal development (such as hyperspecialization and delayed development). The final model is:

$$\frac{dX_i}{dt} = X_i(\sum_{j=1}^{V} \mu_{ij} R_j \overbrace{(1 - \frac{X_i}{K_i})}^{\text{Logistic growth}} - d_i) + \overbrace{\sum_{l=1}^{W} M_{il} X_i X_l / K_i}^{\text{Mutualism}}$$

$$\frac{dR_j}{dt} = r(r_{max} - R_j) - R_j \sum_{i=1}^{W} b_{ij} \begin{cases} \frac{dX_i}{dt}, & \text{if growth-dependent depletion} \\ X_i, & \text{otherwise} \end{cases} \tag{6.3}$$

6.3.1.4 The Wiring of Intelligence

A limitation of mutualism models is that only the activation of nodes is updated. The weight and structure of the network are fixed. While this may be sufficient to explain some developmental phenomena, it is ultimately unsatisfactory. The links themselves should be adaptable, as in the learning of neural networks. An example of learning in the form of updating weights is presented in section 6.3, on the Ising attitude model.

Cognitive growth is a process in which new nodes and links are added during development.

Savi et al. (2019) consider the case where both nodes and links are updated. For example, new facts ($1 + 1 = 2$) and procedures (addition) are developed in the process of learning arithmetic. Links between these nodes may prevent forgetting. We use the Fortuin–Kasteleyn model, a generalization of the Ising model, in which both nodes and links are random variables. An important property of the model is that whenever two abilities are connected, they are necessarily in the same state—that is, they are either both present or both absent. It provides a parsimonious explanation of the positive manifold and hierarchical factor structure of intelligence. The dynamical variant suggests an explanation for the Matthew effect, that is, the increase in individual differences in ability over the course of development.

However, it is difficult to create a growing network with Fortuin–Kasteleyn properties. A simple example of this problem is the random network. In random networks, there is a uniform probability that two nodes are connected. But if we add new nodes to such a network and connect them to existing nodes with the same probability, the existing nodes will have more connections on average. Thus, adding new nodes destroys the uniform randomness of the network; that is, the probability that two nodes are connected is not uniform over nodes anymore. Such a network is a non-equilibrium network (Dorogovtsev and Mendes 2002). Rewiring algorithms to achieve equilibrium exist, but they are not trivial.

6.3.2 SYMPTOM NETWORKS

Mental disorders can be understood as alternative stable states of highly interconnected networks of symptoms.

In the network perspective on psychopathology, a mental disorder can be viewed as a system of interacting symptoms (fig. 6.9). Network theory conceptualizes mental disorders as complex networks of symptoms that interact through feedback loops to create a self-sustaining syndromic constellation (Borsboom 2017).

Like the mutualism model, this is an alternative to the common cause view. Depression could be caused by some malfunction in the brain, a dysregulation of hormones, or even a genetic defect. But, as with general intelligence, no such common cause has yet been found. Drugs work to

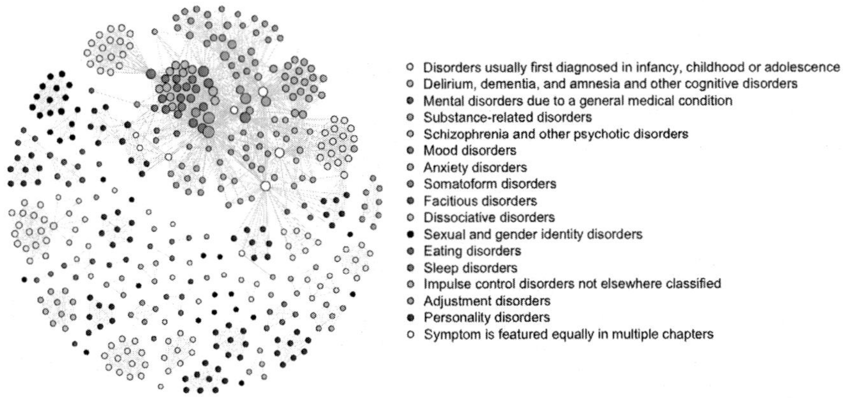

Figure 6.9. The small world of psychopathology. Symptoms are represented as nodes and connected by an edge whenever they figure in the same disorder. Refer to the online version of the book for a color image. *(Adapted from Borsboom et al. (2011) with permission.)*

some extent, but so do most interventions, even placebos and waiting lists (Posternak and Miller 2001). We explicitly offered the network approach as an alternative to the *p* factor account of psychopathology (van Bork et al. 2017).[3] It is called the *p* factor because it is thought to be conceptually parallel to the *g* factor of general intelligence (Caspi et al. 2014). And, again, no one seems to know what *p* might be.

This lack of theoretical progress encouraged the development of network theory (Cramer et al. 2010; Cramer et al. 2016). As mentioned in the introduction of this chapter, this line of research has become popular. Most of this work consists of data analytic studies. In the simplest case, a questionnaire asking about the severity of symptoms is administered to a group of people, sometimes patients, sometimes a mixture of people who do and do not suffer from a disorder. A variety of psychometric approaches, discussed later in this chapter, are used to fit networks to the data. In this way, one learns to understand the structure of psychopathological networks. It is possible to model comorbidity in this way (Cramer et al. 2010; Jones, Ma, and McNally 2021). In the case of major depression and generalized anxiety disorder, sleep problems seem to be a typical bridge symptom (Blanken et al. 2018).

Comorbidity is modeled by bridging symptoms between network clusters.

[3] The original title of this paper was "No Reason to *p*," but the editor did not think it was funny.

The most popular application is to detect which symptoms are central to a disorder (Fried et al. 2016). However, centrality analysis based on cross-sectional data has its limitations (Bringmann et al. 2019; Spiller et al. 2020). This is one reason to focus on individual networks using time-series data, often obtained in experience sampling methods. Again, these techniques are still under development and not without problems (Dablander and Hinne 2019; Haslbeck and Ryan 2022). For a review of the network approach to psychopathology, see Robinaugh et al. (2020).

In terms of building actual models, not as much work has been done. In Cramer et al. (2016), we proposed an Ising-type model, with node values of 0 and 1, representing symptoms being on or off. Nodes were turned on and off based on a probability computed with a logistic function $P = 1/(1 + e^{b_i - A_i^t})$. A_i^t equals the sum of the weighted input from other connected nodes, and b_i is a node-specific threshold that normally keeps nodes in the 0 state. A strong point of this model is that the connections and thresholds were estimated from data. This model is the origin of the connectivity hypothesis. High connectivity within a network of symptoms could lead to a more persistent and severe disorder (for a discussion, see Elovainio et al. 2021).

The connectivity hypothesis suggests that the strength of these connections may lead to the development and maintenance of the disorder.

Since thresholds are generally negative (the 0 state of nodes is the default state), sufficient connectivity is required to have a depression as an alternative stable state. A limitation of this model is that although it is related to the Ising model, the exact dynamics are not well understood.

A similar approach was used by Lunansky et al. (2022) in order to define resilience and evaluate intervention targets. You can see this in the NetLogo "Vulnerability to Depression" model (see fig. 6.10). Another relevant network modeling approach, based on causal loop diagrams, is proposed in Wittenborn et al. (2016).

Resilience is the ability of a system to recover from perturbations and maintain its current equilibrium.

The connection to resilience is interesting. In dynamic terms, resilience is associated not with the healthy or unhealthy state but with the stability of these states (Kalisch et al. 2019). In section 3.3 the less deep minimum is called the metastable state. These states have less resilience than the globally stable state (fig. 6.11).

This suggests a distinction between perturbations and interventions. With interventions, we change the equilibrium landscape to allow a sustainable change to a healthy state. Perturbations (a brief intervention or

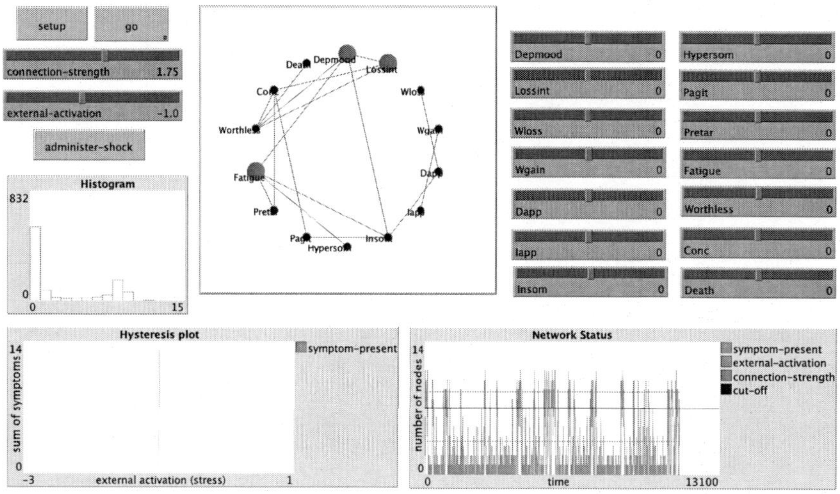

Figure 6.10. The Vulnerability to Depression model in NetLogo. Refer to the online version of the book for a color image.

a positive or negative event) can have a permanent or temporary effect, depending on which state is more resilient. In the situation shown in the top panel of figure 6.11, any perturbation, whether it is a treatment or an alternative (or even being on the waiting list), will work. In the situation on the bottom left, no intervention would have a lasting effect. This analysis of resilience may help to understand the inconsistent results of studies of intervention effects. Monitoring the resilience of the unhealthy state (with catastrophe flags such as anomalous variance) may also be important for timing interventions (Hayes and Andrews 2020). Failed interventions, such as an attempt to quit smoking, are likely to reinforce the unhealthy state (Vangeli et al. 2011).

6.3.3 ISING ATTITUDE MODEL

The network approach has been applied to many other domains outside of intelligence research and the study of psychopathology. Examples include emotion (Lange and Zickfeld 2021; Treur 2019), personality (Costantini et al. 2015; Cramer et al. 2012), interest (Sachisthal et al. 2019), deviations of rational choice (Kruis et al. 2020), and organizational behavior (Lowery, Clark, and Carter 2021). One area where it has been developed into a new theory is attitude research.

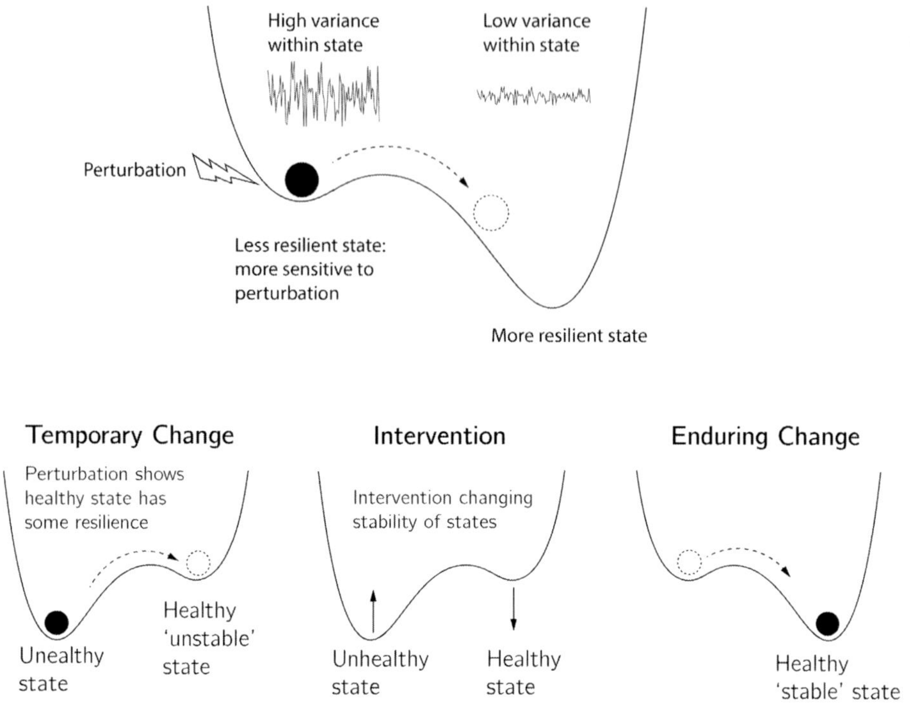

Figure 6.11. Resilience from a complex-systems perspective. If the system is in a less resilient, metastable state, any perturbation will be effective. A perturbation to a metastable state will not last. Lasting interventions change the dynamic landscape of the system.

People have many attitudes—about food, politics, other people, horror movies, the police, etc. They help us make decisions and guide our behavior. Attitudes can be very stable and multifaceted, but they can also be inconsistent and inconsequential. Social psychology has studied attitudes for a long time, and many insights and theories have been developed.

Attitudes are complex constructs. Typical phenomena, such as cognitive dissonance, imbalance, ambivalence, and political polarization, can be well described by a network model.

The formalization of attitude theories has been dominated by the connectionist account (Monroe and Read 2008; Van Overwalle and Siebler 2005). In connectionist models, developed in the parallel distributed processing (PDP) framework, attitude units (e.g., beliefs) form a connected network whose activations (usually between -1 and 1) are updated based on the weighted sum of internal inputs from other units and an external input. These weights or connections are updated according to either the delta rule (a supervised learning rule based on the difference

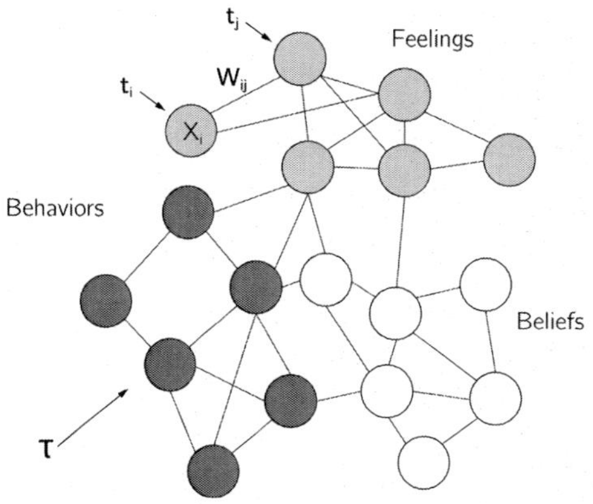

Figure 6.12. The Ising attitude network model. Feelings, beliefs, and behaviors toward an attitude object align when attended to. Nodes are also influenced by local fields, t_i, and a global external field, τ. The connections are weighted.

between the produced and expected output of the network) or the Hebb rule. With this setup, these models can explain a number of phenomena in attitude research. Another network account has been put forward in sociology (DellaPosta 2020).

In this section, I will discuss our network approach to attitudes using the Ising model, which was developed in a series of recent papers. The advantages of this model over the connectionist PDP models are that it is derived from basic assumptions, is better understood mathematically, is easy to simulate, provides a psychological interpretation of the temperature parameter, and can be fitted to data (Dalege et al. 2017).

The Ising model was developed as an alternative to the tripartite factor model of attitudes, in which the attitude, a latent factor, consists of lower-order cognitive, affective, and behavioral factors that each explain observed responses, similar to the Cattell–Horn–Carroll model of general intelligence. The causal attitude model (Dalege et al. 2016) maintains this distinction in cognitive, affective, and behavioral components, but now conceptualizes them as clusters within a network. Nodes represent single feelings, beliefs, and behaviors. In Dalege et al. (2018), this network model is formalized in the form of an Ising model with attention as the equivalent of (the inverse of) temperature. That is, high attention "freezes" the network and leads to consistent and stable positive or negative states of the attitude (the "mere thought effect").

Attitudes are networks of feelings, beliefs, and behaviors toward an attitude object.

Attention is equated to (inverse) temperature.

6.3.3.1 Model Setup

The basic assumptions of the Ising attitude model are that nodes are binary (e.g., one eats red meat or not), that nodes influence each other causally, and that they have specific thresholds (as in the model for depression). An external field (a campaign to eat less meat) could also affect the nodes. The alignment of nodes to other nodes and to the external field depends on one's attention, A, to the attitude object.

Given these simplifying assumptions, which can be relaxed in various ways, we arrive at the random field Ising model (Fytas et al. 2018). This model is not too different from the Ising model described in chapter 5, section 5.2, except that the first term now has two components, a general external effect (τ) and an effect of node-specific (t_i) thresholds ("I just really like the taste of chicken"). The random field Ising attitude model can then be defined as:

$$H\left(\mathbf{x}\right) = -\sum_i^n \left(\tau + t_i\right)x_i - \sum_{<ij>} W_{ij}x_ix_j, \tag{6.4}$$

$$P\left(\mathbf{X} = \mathbf{x}\right) = \frac{\exp(-AH(\mathbf{x}))}{Z} \quad . \tag{6.5}$$

Another difference from the original Ising model introduced in chapter 5 is that the interactions are now weighted and can even be negative. The main technical problem is the same. To compute the probability of a state, one has to compute Z, which is $\sum_{<\mathbf{x}>} exp(-AH\left(\mathbf{x}\right))$, that is, a sum over all possible states (2^n). For large values of n, this is not feasible. One solution is to take a random initial state and use Glauber dynamics to update the states until an equilibrium state is reached. The Glauber algorithm does not require Z. There are faster but less intuitive algorithms, the most popular being the Metropolis–Hastings algorithm, which slightly modifies the Glauber dynamics presented in chapter 5, equation 5.3.

As discussed in chapter 5 (sec. 5.2), another approach to understanding the dynamics of Ising-type models is the mean-field approximation. This requires the assumption that the network is fully and uniformly connected with equal thresholds (known as the Curie–Weiss model). In this approximation $W_{ij} = c$ (all equal) and x_j are replaced by their mean values, which greatly simplifies the energy function. It can be shown that

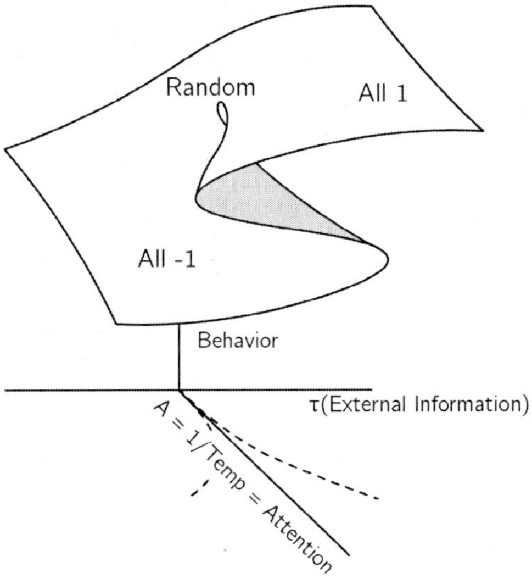

Figure 6.13. The mean field approximation of the Ising attitude model is the cusp. Attention is the psychological equivalent of inverse temperature. Information varies from negative (contra) to positive (pro).

the dynamics of the simple fully connected Ising model are well approximated by the cusp, with the external field as normal and the inverse temperature as the splitting variable.

This is an important result because it makes the use of the cusp in attitude research (see fig. 3.13) less phenomenological. The cusp is now derived from more basic principles (fig. 6.13). Note that here we use attention as the splitting variable, whereas in chapter 3 we used involvement. These are closely related concepts, the difference being the time scale. Attention can change in seconds or minutes, whereas involvement can change in weeks or months. I will use attention and involvement interchangeably.

This mean-field approximation is very robust. In van der Maas, Dalege, and Waldorp (2020), we show via simulation that networks with fewer connections and a distribution of weights, some of which are negative, are still well described by the cusp. This can be easily checked with some R code or in NetLogo. We will make use of the IsingSampler package in R.

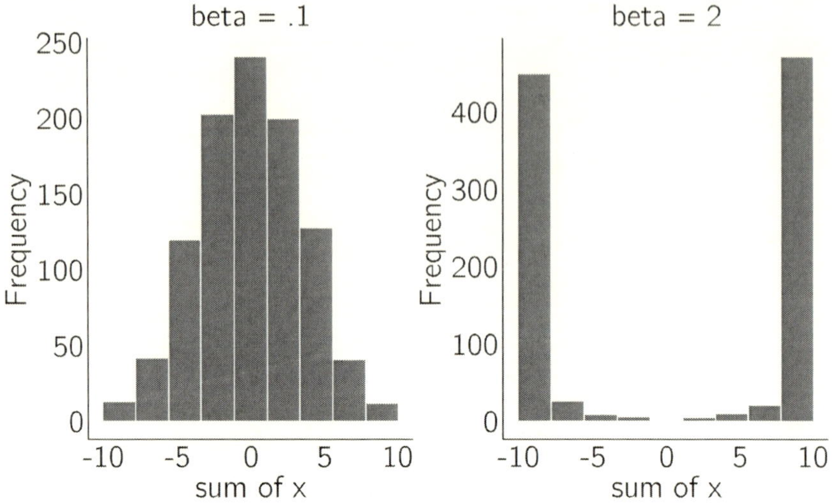

Figure 6.14. The equilibrium distribution of the attitude values (sum of node values) at low and high attention, respectively. This simple simulation demonstrates the mere thought effect (Tesser 1978).

6.3.3.2 Simulation

The IsingSampler function runs the Metropolis–Hastings algorithm *nIter* times and returns the last state. It can return multiple final states for N runs. As input, it takes a matrix of links (**W**), which for the Curie–Weiss model should be constant with 0s on the diagonal. The `thresholds` for each node should be equal. `Beta` , originally the inverse of the temperature ($1/T$), represents attention. The effect of varying `beta` (attention) is shown in figure 6.9.

```
library("IsingSampler")
n <- 10 # nodes
W <- matrix(.1, n, n); diag(W) <- 0
tau <- 0
N <- 1000 # replications
thresholds <- rep(tau, n)
layout(t(1:2))
data <- IsingSampler(N, W, nIter = 100, thresholds,
                     beta = .1, responses = c(-1, 1))
hist(apply(data,1,sum), main = "beta = .1", xlab= 'sum of x')
data <- IsingSampler(N, W, nIter = 100, thresholds,
                     beta = 2, responses = c(-1, 1))
hist(apply(data,1,sum), main = "beta = 2", xlab = 'sum of x')
```

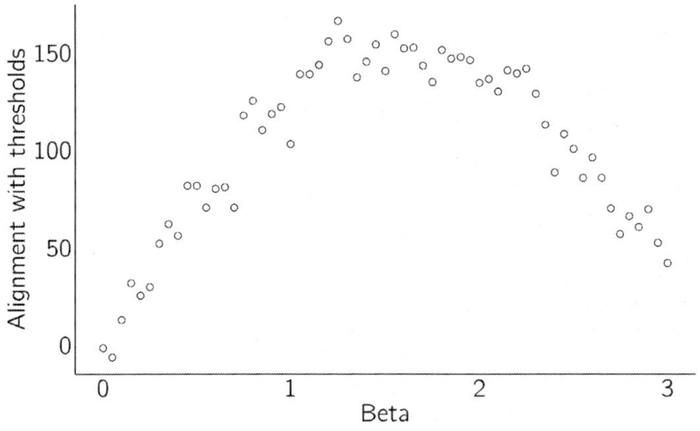

Figure 6.15. At low levels of attention (but not too low), the node values are determined by the thresholds. At higher levels of attention, they are overridden by the collective effect of other nodes. This may explain the difference between implicit and explicit attitude measures.

In Dalege and van der Maas (2020), we simulated the difference between implicit and explicit measures of attitude. The idea is that the individual thresholds contain information about the attitude that can only be detected when attention is moderately low. When attention is too high, the alignment between the nodes dominates the thresholds (fig. 6.15). Indeed, in implicit (indirect) measures of attitude, attention is much lower than in explicit measures such as an interview. This can be simulated as follows:

```
layout(1)
N <- 400; n <- 10
W <- matrix(.1, n, n); diag(W) <- 0
# a random pattern of thresholds:
thresholds <- sample(c(-.2, .2), n, replace = TRUE)
dat <- numeric(0)
beta.range <- seq(0, 3, by = .05)
for(beta in beta.range){
data <- IsingSampler(N, W, nIter = 100, thresholds,
                     beta = beta, responses = c(-1, 1))
# measure of alignment:
dat <- c(dat, sum(thresholds * apply(data, 2, sum)))
}
plot(beta.range, data, xlab = 'beta',
     ylab = 'alignment with thresholds', bty = 'n')
```

We see that for medium attention, the agreement with the thresholds is highest. When attention is 0 or very low, nodes behave randomly and do not correlate with the thresholds. When attention is very high, the effects of node-specific thresholds are masked by the collective effects of other nodes. The principal problem of implicit measurement is that for low to medium attention, the network is quite noisy and measurement reliability will be low. This is why this paper is called "Accurate by Being Noisy."

6.3.3.3 Learning

The connectionist attitude models are capable of "learning," that is, adjusting the weights. This can also be done in the Ising attitude model by using Hebbian learning. Hebbian learning, or "what fires together, wires together," can be formulated as:

$$\Delta W_{ij} = \epsilon (1 - |W_{ij}|) x_i x_j - \lambda W_{ij}, \tag{6.6}$$

which defines the change in weights. Weights will grow to 1 if the nodes they connect are consistently either both 1 or both -1. If they consistently differ in value, the weight grows to -1. If the nodes behave inconsistently, the weight shrinks to 0, due to the last term.

In R, this can be implemented as follows:

```
library(qgraph)
hamiltonian <- function(x, n, t, w){
     -sum(t * x) - sum(w * x %*% t(x)/2)
}
glauber_step  <-  function(x,n,t,w,beta){
  i <- sample(1:n, size = 1) # take a random node
  # construct new state with flipped node:
  x_new <- x; x_new[i] <- x_new[i] * -1
  # update probability
  p <- 1/(1 + exp(beta * (hamiltonian(x_new, n, t, w) -
                          hamiltonian(x, n, t, w))))
  if(runif(1) < p) x <- x_new # update state
  return(x)
}
```

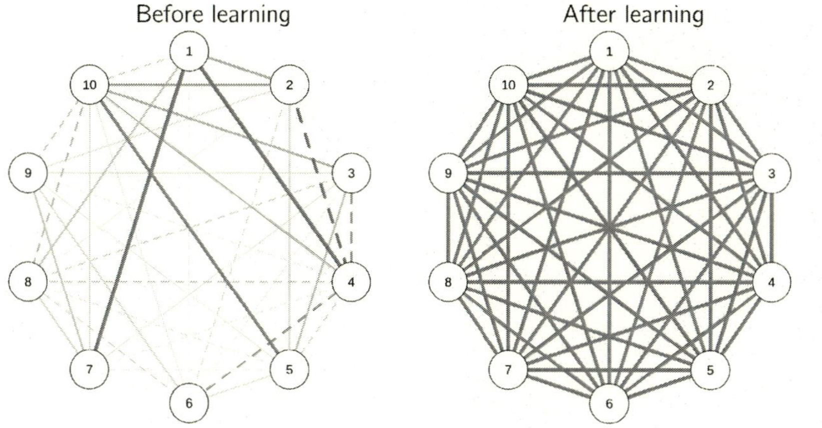

Figure 6.16. Through Hebbian learning, a random (unbalanced) network becomes balanced. Refer to the online version of the book for a color image.

```
layout(t(1:2))
epsilon <- .002; lambda <- .002 # low values = slow time scale
n <- 10; W <- matrix(rnorm(n^2, .0, .4), n, n)
W <- (W + t(W)) / 2 # make symmetric
diag(W) <- 0
qgraph(W); title('before learning')
thresholds <- rep(.2, n)
x <- sample(c(-1, 1), n, replace = TRUE)
for(i in 1:500)
{
x <- glauber_step(x, n, thresholds , W, beta = 2)
# Hebbian learning:
W <- W + epsilon * (1 - abs(W)) * outer(x, x, "*") - lambda * W
diag(W) <- 0
}
# label switching (scale all nodes to positive):
W <- x * t(x * W); x <- x * x
qgraph(W); title('after learning')
```

Figure 6.16 shows the result of this simulation.

In this case we want to update the nodes values using the Glauber dynamics (equation 5.3), which use the computation of the energy of a particular state. Both functions (`glauber_step()` and `hamiltonian()`) are added to the R code.

Due to Hebbian learning, a network evolves from an unbalanced network (random connections) to a consistently balanced network. Without

Developing strong, balanced attitudinal networks has a clear advantage: they do not require much attention to be consistent.

learning, we need high attention to make the attitude network behave consistently. In the learning Ising attitude model (LIAM), weights increase during periods of high attention. The advantage is that in later instances, less attention is required for consistent network behavior (Smal, Dalege, and Maas submitted). In this way we can develop stable attitudes that do not require much attention to be consistent.

6.3.3.4 The Stability of Attitudes and Entropy Measures

Gibbs entropy describes the probability distribution over the different microstates $\mathbf{x}$.

To quantify the consistency of attitudes, we can compute the Gibbs entropy (proposition I.2 in Dalege et al. 2018). The Boltzmann entropy was defined in section 5.2 as the log of the number of ways (W) a particular macrostate can be realized. It measures the inconsistency of a particular attitude state (proposition I.1 in Dalege et al. 2018).[4] Gibbs entropy is more general in that it does not assume that each microstate is equally probable. It is defined as:

$$-\sum_{<\mathbf{x}>} P(\mathbf{x}) \ln P(\mathbf{x}). \tag{6.7}$$

Note that we sum over all microstates (2^n). For small networks, this measure can be computed using the `IsingEntrophy()` function of the IsingSampler package. There is much more to say about the different entropy measures. For instance, Shannon entropy (a measure in information theory) and Gibbs entropy have the same mathematical definition but are derived from completely different lines of reasoning in different fields of science. An introduction to the discussion on entropy measures can be found at the Entropy page of Wikipedia.

6.3.3.5 Tricriticality

A new direction of research concerns Ising-type models with trichotomic node values $(-1, 0, 1)$. In physics this case is known as the tricritical Ising model or the Blume–Capel model (Saul, Wortis, and Stauffer 1974). In physics, the states -1 and 1 could represent the spin of a particle pointing up or down, while 0 could represent a nonmagnetic or spinless state. In

[4] Assuming that all the attitude states (items) are re-encoded as positive (or negative) valued items.

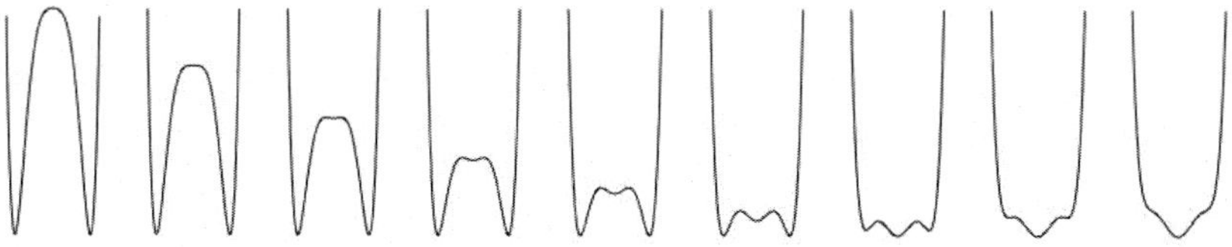

Figure 6.17. The butterfly catastrophe associated with the tricritical Ising model. The potential function can have three minima. In the figure $a = c = 0, d = 5, andb$ varies from 1 to -7.

an attitude model, -1 and 1 may represent pro and con beliefs, while 0 represents a neutral belief. The Hamiltonian of the model includes a penalty for the -1 and 1 states:

$$H(\mathbf{x}) = -\sum_i^n \tau x_i - \sum_{<i,j>} x_i x_j + D \sum_i^n {x_i}^2. \quad (6.8)$$

You can compare this to equation 5.1. The last term penalizes (increases the energy) of the -1 and 1 states relative to the 0 state.

The dynamics of this model are more complicated. It resembles the butterfly catastrophe (Dattagupta 1981), which has a tricritical point. The potential function, $V(X) = -aX - \frac{1}{2}bX^2 - \frac{1}{3}cX^3 - \frac{1}{4}dX^4 + \frac{1}{6}X^6$, has three stable fixed points for specific combinations of values of parameters (see sec. 3.3 and the exercise about the butterfly catastrophe in that chapter). This is relevant to the modeling of attitudes because it opens up the possibility of involved stable in-between attitude positions (see fig. 6.17). The Ising attitude model excludes this.

In the Ising attitude model highly involved persons always radicalize, but more advanced spin models allow for involved nonpartisan positions.

6.4 Psychometric Network Techniques

So far, we have seen examples of theoretical psychological network models in the fields of cognitive, clinical, and social psychology. However, much of the popularity of such models is due to the psychometric approach developed to analyze data using networks. In the last fifteen years, a family of statistical approaches has been developed for all kinds of data and empirical settings. Our psychosystems group (psychosystems.org) has published a book called *Network Psychometrics with R: A Guide for Behavioral and Social Scientists* (Isvoranu et al. 2022). This resource is highly recommended. I will limit myself to a brief overview and some practical examples related to the models presented in the first part of this chapter.

6.4.1 MAIN TECHNIQUES

An important aspect of *Network Psychometrics* is visualizing the network—the process of creating visual representations of the network structure. This helps in interpreting the data. The main R packages for visualization are igraph and qgraph. Both packages include many useful functions.

Network estimation uses statistical methods to estimate the network structure, including which nodes are connected to each other and the strength of those connections.

A more advanced application of network psychometrics is network estimation. The most widely used methods for network estimation are Gaussian graphical models (packages bgms, BDgraph, ggm, psychonetrics, qgraph, BGGM, huge), partial correlation networks (qgraph, qgraphicalmodels), and Ising models (IsingFit, IsingSampler, rbinnet). The mgm package can be used to fit mixed graphical models with a mixture of categorical and continuous valued nodes. The bgms package applies Bayesian estimation and allows testing for missing links. The huge package is used to represent the conditional dependence structure among many variables and is particularly useful when the number of variables is much larger than the sample size.

The estimation is usually followed by a centrality analysis. The most important nodes in the network are identified based on their degree of centrality, which measures the extent to which a node is connected to other nodes in the network. Centrality measures include degree centrality, betweenness centrality, bridge centrality, and eigenvector centrality, among others (packages psych, networktools).

Network comparison is the process of comparing the structure of two or more networks to determine if they are significantly different from each other.

Another important step is network comparison. This can be done using techniques such as the network permutation test, bootstrapping, and moderation analysis (R packages bootnet and NetworkComparisonTest).

Time-series network inference involves analyzing sequential data to identify patterns of interactions and dependencies among variables.

We can perform network inference, inferring causal relationships between nodes, when we have time-series data or when we have intervened in the network. Depending on the type of time series ($N = 1$ time series, $N > 1$ time series, panel data), different modeling options and packages are available (packages psychonetrics, mgm, graphicalVar, mlVar). GVAR returns a temporal network, which is a directed network of temporal relationships, and a contemporaneous network, which is an undirected network of associations between the variables within the same time frame after controlling for temporal relationships.

For a detailed discussion of the reasons for using certain techniques, I again refer you to our recent book. The brief overview I have provided here may soon be obsolete. The CRAN Task View: Psychometric Models and Methods will give you an up-to-date overview.[5] Another option is to use JASP.[6] JASP is a free, open-source statistical analysis program developed under the supervision of Eric Jan Wagenmakers (Huth et al. 2023; Love et al. 2019). It is a user-friendly interface for accessing R packages. Many of the network R packages mentioned above are available in JASP.[7]

All major statistical analyses, both frequentist and Bayesian, are available in JASP.

Finally, I mention semantic network analysis again. A review of statistical approaches (available in R) is provided by Christensen and Kenett (2021).

6.4.2 FITTING THE MUTUALISM MODEL

In section 6.3, I provided code to simulate data. These data can be fitted using JASP. By rerunning the previous code and adding

```
write.table(file= 'mutualism.txt', data, ro = FALSE, sep = '\t')
```

we have a data file ready to analyze in JASP.

After opening this file in JASP, you will see the data. In the Network (Frequentist) tab, select all variables and the EBICglasso option. EBICglasso is an R function from the qgraph package. It calculates the Gaussian graphical model and applies the LASSO regularization to shrink the estimates of links to 0 (Friedman, Hastie, and Tibshirani 2008). This prevents the presence of many irrelevant links without losing predictive value. Alternatively, one could use significant testing or a Bayesian procedure. In JASP, one could use the partial correlation option. I recommended playing around with some options and inspecting additional plots.

Mutualism is an alternative explanation for the positive manifold, which means that the fit of a factor model to such data does not prove that the g factor theory is correct. It can be shown (van der Maas et al. 2006) that the simple mutualism factor model with all $M_{ij} = c$, is equivalent to

[5] https://cran.r-project.org/web/views/Psychometrics.htm

[6] https://jasp-stats.org

[7] You may want to start by reading the blog post on doing network analysis in JASP (https://jasp-stats.org/2018/03/20/perform-network-analysis-jasp/).

the one-factor model. This can be tested in JASP by fitting an exploratory one-factor model to the simulated data. The factor loadings should all be very similar.

As discussed, cross-sectional networks do not provide information about the direction of effects. We can illustrate this as follows.

```
M[,1] <- .2 # strong influence of X1 on all others
M[2,] <- .2 # strong influence on X2 by all others
M[diag(nr_var) == 1] <- 0 # set diagonal of m to 0
```

If we rerun the code and create a centrality plot in JASP, we will see the risks of centrality analysis in cross-sectional networks. Node 2 is the most central, but we know from the simulation that this is because it is influenced by all the others. The node with the most causal power (node 1) does not turn out to be an important central node. With time-series data, we can estimate the direction of the effects. We do this in R:

```
library("graphicalVAR")
# make time series for one persons with some stochastic effects:
data <- run(odes = mutualism, tmax = 1000, table = TRUE,
            timeplot = (i == 1), legend = FALSE,
            after = "state <- state + rnorm(nr_var,
            mean = 0, sd =1 ); state[state < 0] = .1")
data <- data[,-1]
colnames(data) <- vars <- paste('X', 1:nr_var,
                                 sep = '', col = '')
fit <- graphicalVAR(data[50:1000,], vars = vars,
                    gamma = 0, nLambda = 5)
plot(fit, "PDC")
centralityPlot(fit$PDC)
```

The results are shown in figure 6.18. Only the time-series approach provides useful information about possible causal effects.

The M-matrix can take different forms. The typical multifactor structure can be achieved with a block structure.

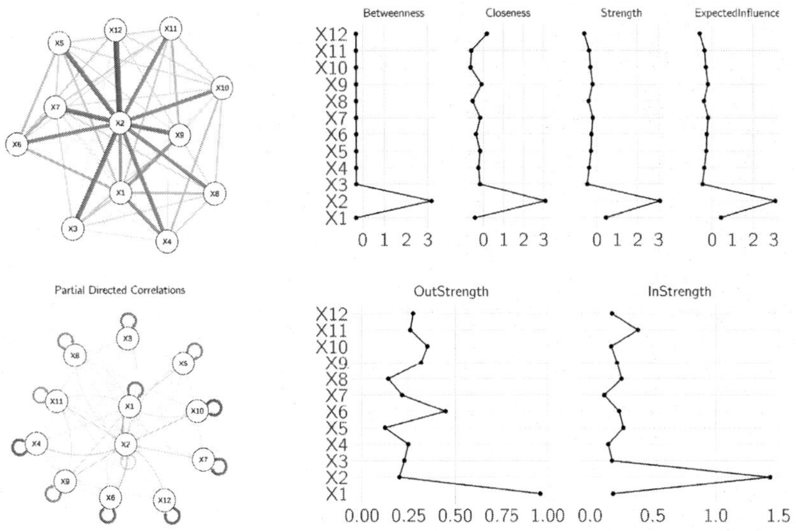

Figure 6.18. The top figure is based on cross-sectional data and incorrectly suggests that node V2 is the most important node. The bottom figure is based on the time series of one individual and correctly shows that V2 is central because it is influenced by other nodes, while V1 is central because it influences other nodes and is therefore more important. Refer to the online version of the book for a more detailed color image.

```
set.seed(1)
factors <- 3
M <- matrix(0, nr_var, nr_var)
low <- .0; high <- .1 # interaction between and within factors
# loop to create M
cat <- cut(1:nr_var, factors)
for(i in 1:nr_var) {
  for (j in 1:nr_var) {
    if (cat[i] == cat[j])
      M[i, j] <- high else M[i, j] <- low
  }
}
M[diag(nr_var) == 1] <- 0 # set diagonal of m to 0
```

In JASP, you can perform network and confirmatory factor analysis. In the latter case, select "3 factors" in the first window and select "assume uncorrelated factors" in the model options. The resulting plots should look like figure 6.19.

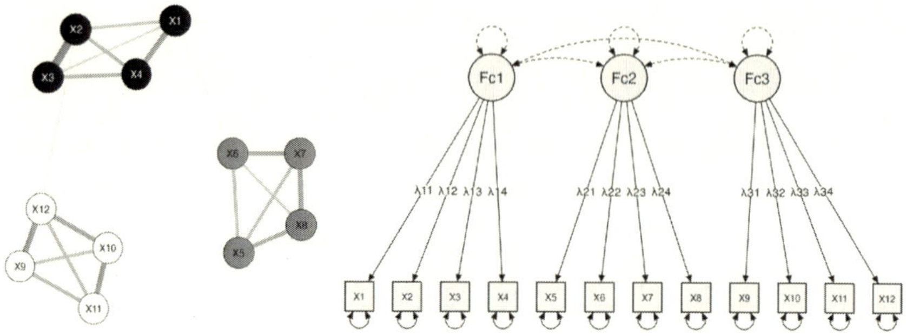

Figure 6.19. The block structure in the mutualism model can be represented as either a network or a factor model. Refer to the online version of the book for a more detailed color image.

6.4.3 FITTING ISING MODELS

With IsingFit we can easily fit cross-sectional data generated with the Ising attitude model. Figure 6.20 shows a good fit of the model. The code for this analysis is:

```
library("IsingSampler"); library("IsingFit")
set.seed(1)
n <- 8
W <- matrix(runif(n^2, 0, 1), n, n); # random positive matrix
W <- W * matrix(sample(0:1, n^2, prob = c(.8, .2),
                replace = TRUE), n, n) # delete 80% of nodes
W <- pmax(W, t(W)) # make symmetric
diag(W) <- 0
ndata <- 5000
thresholds <- rnorm(n, -1, .5)
data <- IsingSampler(ndata, W, thresholds, beta = 1)
fit <- IsingFit(data, family = 'binomial', plot = FALSE)
layout(t(1:3))
qgraph(W, fade = FALSE);
title("Original network", cex.main = 2)
qgraph(fit$weiadj, fade = FALSE);
title("Estimated network", cex.main = 2)
plot(thresholds, type ='p', bty = 'n', xlab = 'node',
    ylab = 'Threshold', cex = 2, cex.lab = 1.5)
lines(fit[[2]], lwd = 2)
```

An empirical example is provided in Dalege et al. (2017). The open-access data ($N = 5728$) come from the American National Election Study of 2012 on evaluative reactions toward Barack Obama. The items and abbreviations are:

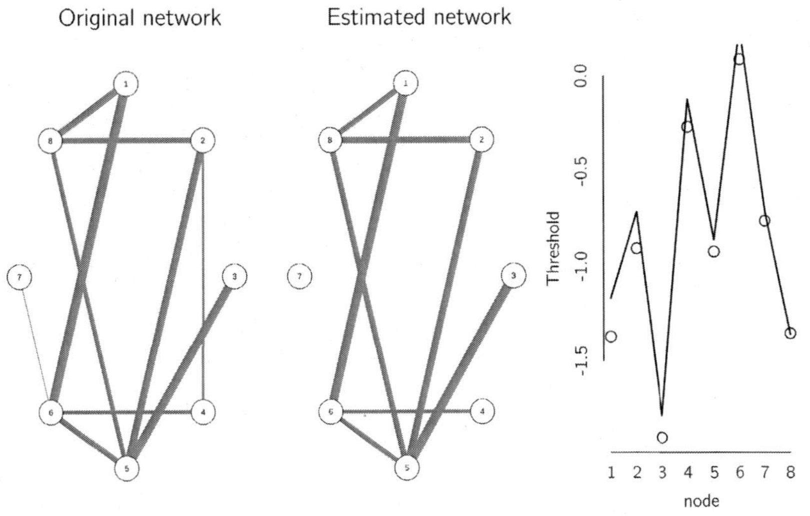

Figure 6.20. The true (original) and the estimated Ising model are in good agreement. The thresholds are also well estimated from the data.

Table 6.1. The abbreviation of items used in figure 6.21.

Items Tapping Beliefs	Abbreviation
"Is moral"	Mor
"Would provide strong leadership"	Led
"Really cares about people like you"	Car
"Is knowledgeable"	Kno
"Is intelligent"	Int
"Is honest"	IIns
Items Tapping Feelings	
"Angry"	Ang
"Hopeful"	Hop
"Afraid of him"	Afr
"Proud"	Prd

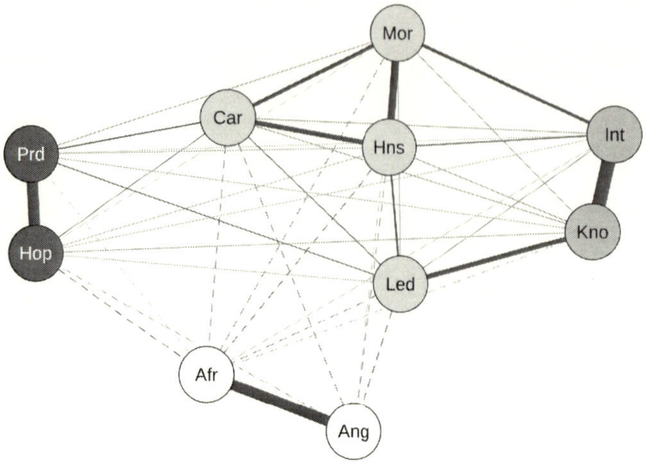

Figure 6.21. The attitude toward Obama in 2012 represented as a network. Refer to the online version of the book for a color image.

We can use Isingfit and add community detection:

```
Obama <- read.table("data/Obama.txt", header = TRUE)
ObamaFit <- IsingFit(Obama, plot = FALSE)
ObamaiGraph<- graph_from_adjacency_matrix(abs(ObamaFit$weiadj),
        'undirected', weighted = TRUE, add.colnames = FALSE)
ObamaCom <- cluster_walktrap(ObamaiGraph)
qgraph(ObamaFit$weiadj, layout = 'spring', cut = .8,
       groups = communities(ObamaCom), legend = FALSE)
```

Figure 6.21 shows the network. The red nodes represent negative feelings toward Barack Obama; the green nodes represent positive feelings; the light blue nodes represent judgments primarily related to interpersonal warmth; and the purple nodes represent judgments related to Obama's competence. This community structure is consistent with the postulate of the Ising attitude model that similar evaluative responses cluster (Dalege et al. 2016). Finnemann et al. (2021) present additional examples and applications of other packages.

6.5 Challenges

A large amount of work has been done since the early work on network psychology, the mutualism model, and the paper on the network perspective on comorbidity. In particular, network psychometrics has taken off in an unprecedented way. One could say that modern psychometrics is being reinvented from a network perspective. For every type of data and research question, a network approach seems to be available. For example, there are R packages for meta-analysis from a network perspective (Salanti et al. 2014). I also note that much has been done to understand the relationship between network psychometrics and more traditional techniques such as item response theory (Marsman et al. 2018), factor models (Waldorp and Marsman 2022), and structural equation modeling (Epskamp, Rhemtulla, and Borsboom 2017). Nevertheless, there are still many challenges for both psychological network modeling and network psychometrics.

6.5.1 PSYCHOLOGICAL NETWORK MODELING

Despite all the work on this, I can only conclude that this theoretical line of research is still in its infancy. The strength of the application to intelligence is that it provides an alternative to the g factor model, which is also nothing more than a sketch of a theory. The extensions of the mutualism model (de Ron et al. 2023; Savi et al. 2019) add new steps, but remain rather limited models. One reason for this state of affairs is that it is really hard to pinpoint the elementary processes involved in intelligence, and indeed in any psychological system.

This is perhaps less of a problem in the factor account because the indicators are interchangeable in a reflective factor model. Once one has a sufficiently broad set of indicators, the common cause estimate will be robust. In a formative model, each indicator contributes a specific meaning to the index variable. However, this is not a reason to prefer the common cause model (van der Maas, Kan, and Borsboom 2014).

The other modeling examples suffer from the same problem. In the clinical psychology models, we define the nodes either as the symptoms specified in the Diagnostic and Statistical Manual of Mental Disorders (DSM) or as the questions asked in interviews or questionnaires, with the advantage that we then have data to fit the model. But, again, we have no real way of knowing the elementary processes in clinical disorders and their interactions (Fried and Cramer 2017).

If we miss important elementary nodes, this may seriously affect the validity of our models and psychometric network analyses.

A way out has been mentioned in the context of the Ising attitude model, using the mean-field approximation. If we are only interested in the global behavior of the attitude (hysteresis, divergence), we can ignore the

In mean-field analyses, the specification of all nodes is less important.

specification of the nodes (another interchangeable argument). But if one wants to intervene on specific nodes or links of a clinically depressed person, this is not sufficient.

Another critical point is that these models increase our understanding of psychological phenomena, but seemingly not our ability to predict or intervene. For example, the Ising attitude model helps us understand the role of involvement (attention) in the dynamics of attitudes. If this factor is too high, persuasion will be extremely difficult due to hysteresis. Anyone who has ever tried to argue with a conspiracy theorist knows what I mean. But too little attention is also a problem. In the model, these are people who are sensitive to the external field, for example, you tell them to clean their room, but as soon as you leave, the attitude falls back into random fluctuations. The message gets through but does not stick. I find this insightful, but I must admit that it does not provide us with interventions. We don't know how to control attention or engagement, although more work can and will be done on this.

For intelligence, the model suggests that the active establishment of near and far transfer might be effective. A disappointing lesson from developmental psychology is that transfer does not always occur automatically (e.g., Sala et al. 2019). However, strategies for improving transfer do exist (Barnett and Ceci 2002) and, according to the mutualism model, should have a general effect.

In chapter 1, section 1.3, I mentioned the case of the shallow lake studied in ecology, where catching the fish was a very effective intervention, while addressing the cause, pollution, was ineffective due to hysteresis. Ecologists now know why this is so and have developed models to explain this phenomenon. However, I did not mention that this intervention was suggested not by modeling work but by owners of ponds who observed that ponds without fish sometimes spontaneously tipped to the clear state. This is not an uncommon path in science, and it may well occur in clinical psychology. The touted extraordinary successes of electroshock therapy for severe depression or new drugs (MDMA) for post-traumatic stress disorder could be our "fish." But these claims have also been criticized (e.g., Borsboom, Cramer, and Kalis 2019; Read and Moncrieff 2022).

Although much more progress can be made in network modeling of psychological systems, it is advisable to be realistic. Progress in mathematical modeling of ecosystems has also been slow. The formalization of psychological models is of interest for many reasons (Borsboom et al. 2021), but will only be effective if we also make progress in other areas, such as measurement (ch. 8).

Ecosystems and human systems are devilishly complex.

6.5.2 PSYCHOMETRIC NETWORK ANALYSIS

This approach is also not without its problems, some of which are related to the problems of psychological network modeling. For example, the definition of nodes and the risk of missing nodes in the data is a serious threat. This problem is not unique to network analysis; simple regression analysis suffers from the same risks. Another common threat to many applications of psychometric network analysis is the reliance on self-report in interviews or questionnaires. Generalizability, which may depend more on the choice of sample and measurement method than on the statistical analysis itself, is another example of a common problem in psychology in general, and psychometric network analysis in particular.

Psychometric network analysis has been criticized because the results are difficult to replicate (Forbes et al. 2017). Replication of advanced statistical analyses, whether structural equation modeling, fMRI, or network psychometrics, is always an issue, but some solutions have been developed (Borsboom et al. 2017, 2018; Burger et al. 2022).

For network analysis, a number of safeguards have been developed to increase replicability

A final important issue concerns causality. Network models estimated from cross-sectional data are descriptive rather than causal; that is, they do not provide information about the direction of causal relationships between variables. Developing methods for inferring causality from network models is an important challenge in the field.

The move to time-series data (either $N = 1$, $N > 1$, or panel data) partially solves this problem (Molenaar 2004). With time-series data, we can establish Granger causality, a weaker form of causality based on the predictive power of one variable over another in a time-series analysis. However, the relationship may be spurious, influenced by other variables. Network analysis on time series often requires a lot of reliable and stationary data. An important issue is the sampling rate of the time series. In general, to accurately estimate a continuous time-varying signal, it is necessary to sample at twice the maximum frequency of the signal. This is called the Nyquist rate. Another issue is the assumption of equidistance between time points

Granger causality suggests a directional influence when one time series predicts another, without necessarily implying a true causal relationship.

(Epskamp et al. 2018), which can be circumvented by using continuous-time models (Voelkle et al. 2012).

While these problems are not unique to network psychometrics, they are serious problems in practice (see Hamaker et al. 2015; Ryan, Bringmann, and Schuurman 2022). The ultimate test of causality requires intervention. For new work along this line, I refer to Dablander and van Bork (2021) and Kossakowski, Waldorp, and van der Maas (2021).

Hopefully, the combination of observational and experimental data can provide sufficient information to properly estimate causal relationships in directed acyclic graphs.

Perhaps we can also think of other ways. A simple, but difficult to implement, procedure is to ask subjects about the links in their networks. If one claims not to eat meat because of its effect on the climate, one might consider adding a directed link to this individual's network (Rosencrans, Zoellner, and Feeny 2021). Deserno et al. (2020) used clinicians' perceptions of causal relationships in autism. These relationships were consistent with those found in self-reported client data. The main problem is that there are many more possible links than nodes to report on, which makes the questionnaires extremely long and tedious to fill out. Brandt (2022) applied conceptual similarity judgments to construct attitude networks. Alternatively, one could try to estimate links from social media data, interviews, or essays using automatic techniques (Peters, Zörgő, and van der Maas 2022).

6.6 Conclusion

Despite these critical remarks, it is safe to say that the psychological network approach has made great progress in a very short time. The mutualism paper was only published in 2006. It is a new and unique application of the complex-systems approach within the field of psychology. For me, the common-cause approach is theoretically unsatisfactory because the common causes are purely hypothetical constructs. For both the g and p factors, there is no reasonable explanation for their origin. In contrast, the reciprocal interactions between function and symptoms that underlie the network theories of intelligence and psychopathology are hardly controversial. It is also good to note that the network approach is not inconsistent with statistical factor analytic work in psychology. It is a matter of interpreting the general factor as a common cause or as an index. The formative interpretation of factors is consistent with the network approach.

In the next chapter, we take the step to modeling social interactions using social networks, where the nodes represent agents that move to new locations, learn a language, share cultural attributes, and have opinions. The focus is on opinion networks. The chapter ends with our own agent-based model of opinion dynamics, which builds on the Ising attitude network model. This opinion model is a social network of within person attitude networks. To simplify we will use the cusp description of the Ising attitude network model at the within-person level. In essence, the model is based on interacting cusps, similar to the model for multifigure multistable perception introduced in chapter 4 (sec. 4.3). This opinion model suggests a new explanation and a new remedy for polarization. ❧

6.7 Exercises

1. Reproduce the degree distribution of the Barabási–Albert model shown on the scale-free network Wikipedia page. Use sample_pa from the igraph library. (*)

2. Open and run the "Preferential Attachment" model in NetLogo. Replace the line `report [one-of both-ends] of one-of links` with `report one-of turtles`. New nodes will now connect to a random node. Does this result in a random network? (*)

3. Make a hysteresis plot in the "Vulnerability to Depression" model in NetLogo (see sec. 6.3). (*)

4. In the "Vulnerability to Depression" model you can deactivate all symptoms at once with the `administer-shock` button. It is as if you give the network an electric shock that resets all the symptoms. Try to find a settings of the `connection-strength` and `external-activation` that creates a disordered network (above the black line in the `network status` plot) whereby administering a shock makes the system healthy again. Is this healthy state long-term stable? (*)

5. Compute the Gibbs entropy for the Learning Ising Attitude model during the learning process (see sec. 6.3). Show in a plot that learning minimizes the Gibbs entropy. (**)

6. Install and open JASP (jasp-stats.org). Open the data library: "6. Factor." Read all the output and add a confirmatory factor analysis. What is the *standardized* factor loading of Residual Pitch in the confirmatory one-factor model? (*)

7. Read the blog "How to Perform a Network Analysis in JASP."[8] Reproduce the top plots of figure 6.18. Generate the data using the R code in section 6.4, import the data into JASP, and perform the network analysis. (*)

8. Study the R code for the case where the M-matrix consists of three blocks (sec. 6.4). Generate the data and import into JASP. Apply exploratory factor analysis, check the fit for one to three factors, and report the p-values. Fit the confirmatory three-factor model. Does it fit? Add V1 to the second instead of the first factor. How do you see the misfit? (*)

9. How can you generate data for a higher-order factor model using the mutualism model? What should be changed in the code of the **M**-matrix for the case of three blocks? Show that the three-factor solution (assuming uncorrelated factors) does not fit the resulting data. Fit a higher-order factor model and report the p-value of the goodness of fit. How does the network plot change? (**)

10. Generate data for a network in a cycle (v1 -> v2 -> v3...v12 -> v1). Fit a network and an exploratory factor model. Does this work? What does this tell us about the relationship between the class of all network models and all factor models? (**)

11. Fit a Bayesian network in JASP to the data generated for figure 6.20. Warning: the GM in JASP expects (0, 1) data. Check that only the simulated links have high Bayes factors. (*)

[8] https://jasp-stats.org/2018/03/20/perform-network-analysis-jasp/

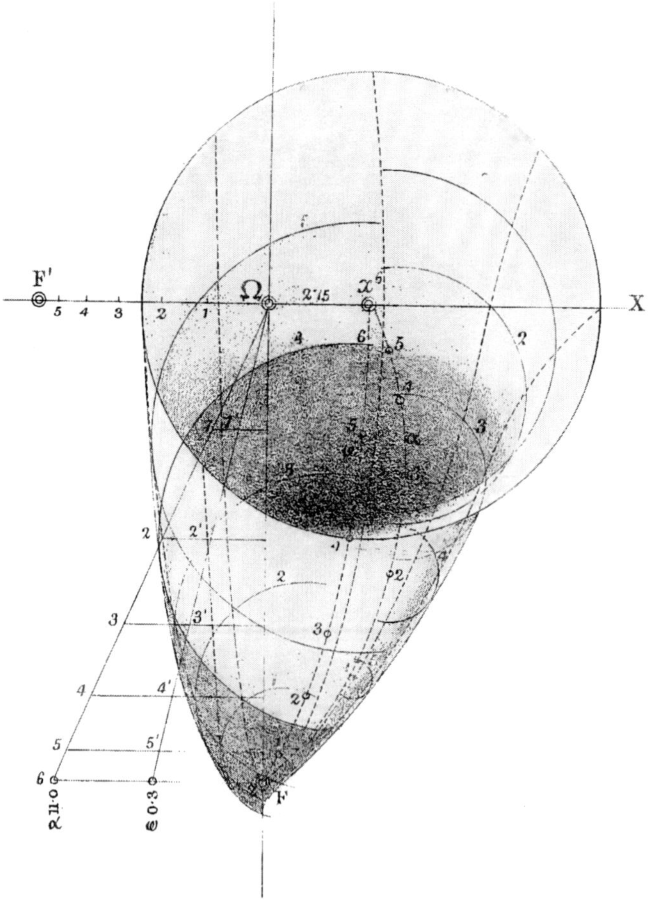
F'
Ω
2·15
X
F

SOCIOPHYSICS

7.1 Introduction

This final chapter is dedicated to the dynamics between people. Networks of social interactions undoubtedly meet the criteria of a complex system and often exhibit unpredictability, sudden changes, and self-organization. Models of the dynamics between people are the domain of social scientists, such as sociologists, but also scientists with a background in statistical physics. Again, it is necessary to simplify, and the victim in this case is psychology. It is a challenge to add a little more psychology to the agents in these models so that more phenomena can be explained without making these models too complex to study. It is very easy to cross that line. The second part of the chapter deals with what I call *psychosocial models*.

RESOURCES
The electronic version of this book, including code blocks and links, can be viewed and accessed at: *santafeinstitute.github.io/ComplexPsych/*

In most sociophysics models, people are reduced to binary nodes, on or off, pro or contra, or to systems with only one continuous attribute.

I use the somewhat controversial term *sociophysics* as a label for this very diverse field of research, which has roots in many different sciences. It is as (in)correct as the generally accepted term *psychophysics* in psychology. What I like about it is that it emphasizes the intention to arrive at a formalized, mathematically stated theory of social processes, often based on formalisms first developed in physics. Alternative labels are *social physics*, *computational social science*, and *agent-based modeling* (Goldstone and Janssen 2005).

In my broad definition of sociophysics, it deals with many different systems and processes. Examples that I have already discussed are the synchronization of movement, such as in the swinging of legs in a two-person system or in a crowd of people fleeing a fire in a building. Granovetter's threshold model of joining a dancing crowd is another example. Many papers deal with the spread of opinions, fashions, rumors, etc. This will be the main topic of this chapter. But this field also includes work on segregation, cooperation, crime, economic systems, and much more. I will present some key examples and then turn to models of the social dynamics of opinions.

Some important follow-up books to read are Smaldino (2023) on modeling social behavior; Bowles and Gintis (2011) on cooperation and altruism; Durlauf and Young (2001) on modeling social economics; Epstein and Axtell (1996), Epstein (2006), and Epstein (2014) on building artificial societies; and Miller and Page (2007) on computation models of social life. There are also some excellent review articles on this area of research (Castellano, Fortunato, and Loreto 2009; Noorazar 2020; Proskurnikov and Tempo 2017; Jusup et al. 2022). A less mathematically focused review is provided by Flache et al. (2017).

7.2 Some Famous Examples

7.2.1 SEGREGATION

The most famous computational model for studying segregation is Schelling's model (Schelling 1971). The simplified version of this model assumes a two-dimensional space, a cellular automaton as in NetLogo, where each location or cell is occupied by one of two types of agents or is empty. The density parameter determines how many locations are occupied. Agents stay at their location if a certain percentage of their eight neighbors, B, are of the same type as the agent. So, if $B = 0\%$, nobody moves. For values of B close to 100%, everybody moves all the time. What Schelling showed with his model was that even low levels of intolerance, near 30%, lead to segregation.

Even if people can accept up to 70% of "foreign" neighbors, segregation still occurs.

The NetLogo model Segregation (Social Science sample model) demonstrates this model. Especially at high densities, transitions between the unsegregated state, the segregated state, and a mixed state where agents keep moving can be seen (Gauvin, Vannimenus, and Nadal 2009). Instead of playing around with the sliders, we can also use the BehavioralSpace option. Figure 7.1 shows the settings for this option, the R code, and the results. Note that without visualizations NetLogo is pretty fast.

One concept we saw in Chapter 5, section 5.4, in the context of Haken's work, was that of the order parameter. The order parameter literally measures the order in the system. We look for an order parameter that captures the qualitative phenomena in the model, such as sudden changes and hysteresis. In the NetLogo simulation, I used the percentage of similar neighbors. Gauvin, Vannimenus, and Nadal (2009) propose two, probably better, order parameters: a segregation coefficient, which requires the identi-

Finding an appropriate order parameter is not always easy, and multiple definitions are often possible.

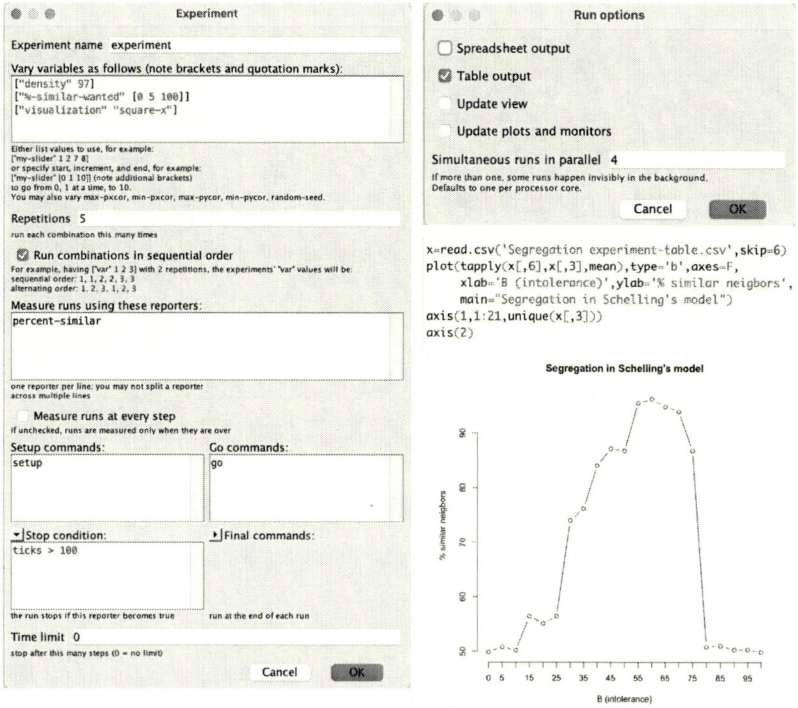

Figure 7.1. BehaviorSpace settings and R code to visualize the effects of intolerance on segregation.

fication of clusters, and the density of unwanted locations, which better distinguishes between the three states of the system.

Clearly, this is only an initial model that can be extended in many ways, some of which are already explored in Schelling's original paper. For an interesting historical sketch of this line of research, see Hegselmann (2017). Another NetLogo model inspired by Schelling's research is called Party. In this model, partygoers experience discomfort and change their groups if their current group consists of a disproportionate number of individuals of the opposite sex. A recent review with a focus on urban dynamics can be found in Jusup et al. (2022).

7.2.2 THE EVOLUTION OF LANGUAGE: THE NAMING GAME

Languages are complex systems that emerge in a self-organizing cultural process (Steels 1995). Steels developed the naming game (fig. 7.2) to study the evolution of language. For a review I refer to Chen and Lou (2019). Castellano, Fortunato, and Loreto (2009) provide a brief review.

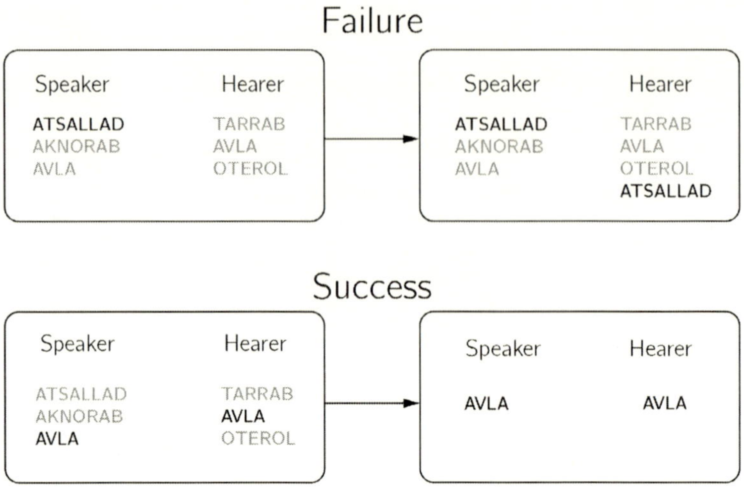

Figure 7.2. The naming game. The speaker chooses the highlighted word. If the listener does not know the word (failure), she adds it to her inventory. In case of success, both agents delete their inventories, keeping only the spoken word. (Adapted from Castellano et al., 2009)

The game begins with blank lists for all agents. In the first round, the speakers invent a new unique word, assuming that the possibilities are endless. We assume that there is only one object to name and that the social network is fully connected. Two agents are randomly selected from a population: one as the "speaker" and the other as the "hearer." The speaker chooses a word from her vocabulary for the object. If the hearer is unfamiliar with the word, she incorporates it into her lexicon. If she recognizes the word, both clear their vocabularies, retaining only the word that was spoken.

The R code to simulate this process is:

```
resample <- function(x, ...) x[sample.int(length(x), ...)]
n <- 1000; iter <- 100000
x <- vector("list", n)
n_possible_words <- 1000000 # should be very high
total_words <- total_unique_words <- numeric(iter)
for(i in 1:iter){
  j <- sample(1:n, 1) # speaker
  k <- sample((1:n)[-i], 1) # hearer
  # if speaker has no words, make one up (actually just number)
  if(length(x[[j]]) == 0)
    x[[j]][1] <- sample(1:n_possible_words, 1) else
    {
     # choose a word (actually just number):
      spoken_word <- resample(x[[j]], 1)
      if(any(x[[k]] %in% spoken_word)) # hearer knows the word
        {
          x[[j]] <- spoken_word # erase list except spoken_word
          x[[k]] <- spoken_word # erase list except spoken_word
        } else # hearer does not know the word
        x[[k]] <- c(x[[k]], spoken_word)  # add word to list
  }
  total_words[i] <- length(unlist(x))
  total_unique_words[i] <- length(unique(unlist(x)))
}
layout(1:2)
plot(total_words, type = 'l', xlab = 'time', bty = 'n')
plot(total_unique_words, type = 'l', xlab = 'time', bty = 'n')
unique(unlist(x)) # winning word
```

which results in figure 7.3.

After a phase in which agents use lots of different words, a language consisting of just a single word emerges abruptly. A simplification would be the case that agents start with either word A, word B, or both words A and B. So no new words are invented. For this case we can write the time evolution in three differential equations:

$$
\begin{aligned}
\frac{dX_A}{dt} &= -X_B X_A + \frac{1}{2} X_{AB} X_{AB} + X_A X_{AB}, \\
\frac{dX_B}{dt} &= -X_A X_B + \frac{1}{2} X_{AB} X_{AB} + X_B X_{AB}, \\
\frac{dX_{AB}}{dt} &= 2 X_A X_B - X_{AB} X_{AB} - (X_A + X_B) X_{AB}.
\end{aligned}
\tag{7.1}
$$

The first equation can be understood as follows: Speakers B talking to listeners A turn listeners A into AB agents. The loss in A agents is

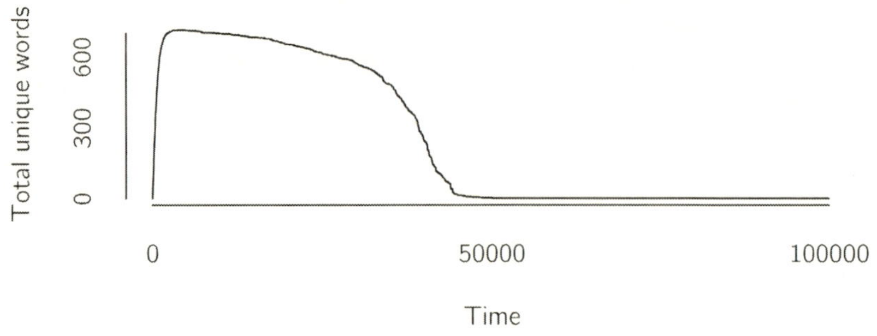

Figure 7.3. The order parameter, the total number of unique words, undergoes a phase transition to a state where everyone uses the same word for the object (compare Castellano et al., 2009, fig.2).

Although extremely simplified, the naming game illustrates that coexistence of languages is difficult in a fully connected network.

$-X_B X_A$. AB agents talking to AB agents become A or B agents, depending on the speaker's random choice of A or B ($X_{AB}X_{AB}/2$). If speaker A talks to an AB agent, the latter becomes an A agent ($X_A X_{AB}$). These three terms together define the change in X_A. The other equation follow the same logic. It is easy to implement this in Grind. There are three possible outcomes: (1) If the initial proportion of A is greater than B, A wins; (2) if the initial proportion of B is greater than A, B wins; and (3) if these proportions are exactly equal, all three options A, B, and AB coexist, but this equilibrium is unstable.

7.2.3 CULTURAL DYNAMICS: THE AXELROD MODEL

Homophily refers to the tendency or preference for individuals to associate or connect with others who are similar to themselves.

Robert Axelrod (1997) introduced an influential model of cultural diffusion based on the effects of social interactions and homophily. In Axelrod's model, individuals become more similar through interactions—but only if they already share some cultural features.

In the model, agents have F cultural features (e.g., beliefs, habits), each of which has Q possible nominal values. Agents are organized in some kind of network, for instance, a fully connected one. At each time step, two agents are randomly selected. We count the number of shared features, such as features with the same value. This number, divided by F, gives the probability that one of the mismatching features of one of the two agents is set equal to that of the other. Thus, if they differ on all features, nothing happens. If 50% of their features have the same value, then one of the mismatching features will change for one of the agents with a probability of 0.5.

The combination of interaction and homophily creates a self-reinforcing dynamic that often leads to global convergence toward a single culture. However, for certain choices of F and Q, the model converges to a state of diversity. In the simulation below, I use the number of remaining cultures as a simple order parameter. Castellano, Marsili, and Vespignani (2000) present detailed analyses of the phase transitions in this model using more advanced order parameters. As you can imagine, this model can be extended in many ways, for instance by introducing ordinal instead of nominal states (Macy, Flache, and Takacs 2006).

The following R code of this model generates the first plot in figure 7.4 . You can play with the values to see different cases.

```
resample <- function(x, ...) x[sample.int(length(x), ...)]
n <- 100; iter <- 50000
F <- 4 # features
Q <- 4 # nominal levels per feature
uniques <- numeric(iter)
x <- matrix(sample(1:Q, replace = TRUE, n * F), n, F)
for(i in 1:iter){
  j <- sample(1:n, 1)          # agent 1
  k <- sample((1:n)[-j], 1)    # agent 2
  w <- sum(x[j,] == x[k,]) / F # agreement
  if(w < 1 & runif(1) < w) {
     # which (unequal) feature to update:
     f <- resample(which(x[j,] != x[k,]), 1)
    x[j,f] <- x[k,f] # update
    }
  uniques[i] <- nrow(unique(x))
}
plot(uniques[1:i], type = 'l', lwd = 2, xlab = 'time',
     ylab = '# unique cultures', bty = 'n', ylim = c(0, 80),
     main = paste0('Axelrod model with F = ', F, ', Q = ', Q))
```

What I find appealing about the Axelrod model is its multidimensionality. Many are more concerned about polarization in the United States than in the Netherlands. In the United States, it seems that all aspects of life are interconnected, and even seemingly unrelated factors such as one's choice of jeans or favorite sport are correlated with one's political beliefs. This greatly limits the possibilities for depolarization.

As long as we have some characteristics or qualities in common with other people on certain dimensions, there is hope for society.

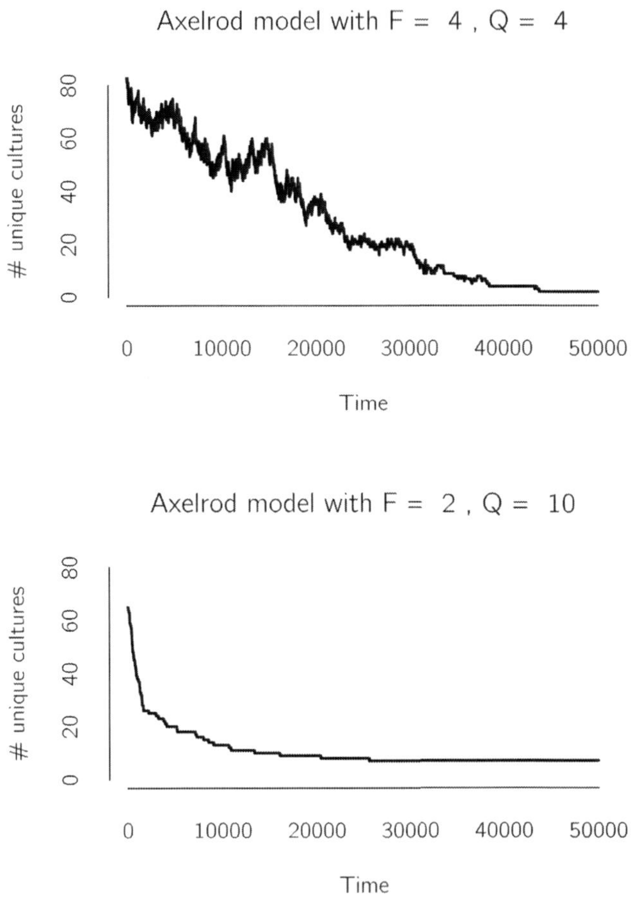

Figure 7.4. Two runs of Axelrod's model. For four features with four possible values, only one culture remains. For the second case ($F = 2, Q = 10$), multiple unique cultures, different in all features, emerge.

7.3 Dynamics of Opinions

In that same famous paper, Axelrod asked an important question: "If people tend to become more alike in their beliefs, attitudes, and behavior when they interact, why don't all such differences eventually disappear?"

The answer to Axelrod's question is usually posed in terms of limited interactions between agents. In Axelrod's model, for example, it was due to selective interaction between agents. In continuous-opinion models, it is due to bounded confidence, that is, agents that are too different refuse to interact. In some models, there is simply no connection between subgroups in a network.

Bounded confidence refers to the concept that individuals are influenced by the opinions of others only when those opinions fall within a certain range of their own opinions.

It is hard to count the number of opinion-spread models, but it could easily be in the hundreds, if you count all the variants. They all share a few building blocks. First, there has to be some topology to the social network. Modelers make different choices here. Often, fully connected networks are assumed because they allow an analytical (mean-field) approach; others use random networks, lattices, small-world networks, etc. The problem is that we don't really know how real social networks work, except that they are incredibly complex (Newman and Park 2003). Second, you need to define some interaction rules. For example, two agents might end up in the middle after a discussion, one agent might copy the other's state, or one agent might take over the majority vote in its local neighborhood. Finally, you have to define opinion. I will first discuss several discrete-opinion models.

A major division in social contagion models is whether opinions are defined as discrete or continuous.

7.3.1 DISCRETE-OPINION MODELS

7.3.1.1 Voter Models

The simplest possible model seems to be the voter model. In its basic form, with only two possible opinions $(-1, 1)$, two connected agents A and B meet, and A simply copies B's opinion. What happens in this simple system depends on the topology of the network, that is, its dimension (either $d = 1$ (on a line), $d = 2$ (a lattice) or $d > 2$) and its size N. In more than two dimensions and with infinite size, the voter system does not converge, but in other cases it converges to a state in which all opinions agree (either all -1 or 1) (Castellano, Fortunato, and Loreto 2009; Redner 2019). How long it takes to converge can also be derived analytically. The convergence time is proportional to N^2, for voters on a line ($d = 1$), $NlnN$ for $d = 2$, and N

for $d > 2$. Thus, the convergence is slowest when agents are connected in a line.

The probability of ending up in the +1 state is equal to the initial probability of +1s.

In the heterogeneous voter model, each agent A copies the opinion of agent B with some probability r_i. In this way, one can study the effect of stubborn voters (with low r_i). It turns out that the small group of stubborn individuals (sometimes called zealots) can overcome the majority opinion. Many other variants have been analyzed, such as adding memory and noise to the voters (Castellano, Fortunato, and Loreto 2009). It is also possible to consider three groups, left, center, and right, where left and right do not interact. In this case, depending on the initial proportion, we end up in a state of full consensus in one of the states or with a mixture of extremists without centrists (Redner 2019). Finally, the topology of the social network plays a role.

In general, consensus is easier reached in scale-free networks with broad degree distributions.

Another approach has been proposed by Martins (2008). The Continuous Opinions and Discrete Actions (CODA) model combines discrete and continuous aspects of opinion dynamics. Agents act discretely but update their continuous opinions based on observations of other agents' discrete actions.

In CODA, there are two choice options, A and B, and agent i has some subjective probability p_i that A is the best option, and $1 - p_i$ for B. The actual choice is made according to $sgn\ (p_i - .5)$, so A is chosen when $p_i > .5$. Next, the agent observes other agents. Agents assume that other agents make rational choices, that is, choose A when A is the best option with a probability a that is larger than .5. In running the model, it is convenient to work with the log-odds of probabilities, $v_i = ln(p_i/(1 - p_i))$. Using Bayes's theorem, we can update v_i to $v_i + a$ when agent j chooses A and to $v_i - a$ if the choice is B. Martins (2008) integrates these decision rules with the voter model, showing extreme forms of polarization, that is, a strongly bimodal distribution of opinions.

7.3.1.2 More Discrete-Opinion Models: Majority Type Models

In the voter and Axelrod models, interactions are limited to two agents. When multiple neighbors have an impact on each agent, many new options arise. One option is the Ising model (Galam, Gefen, and Shapir 1982). Agents switch sides with a probability that depends on the states of their

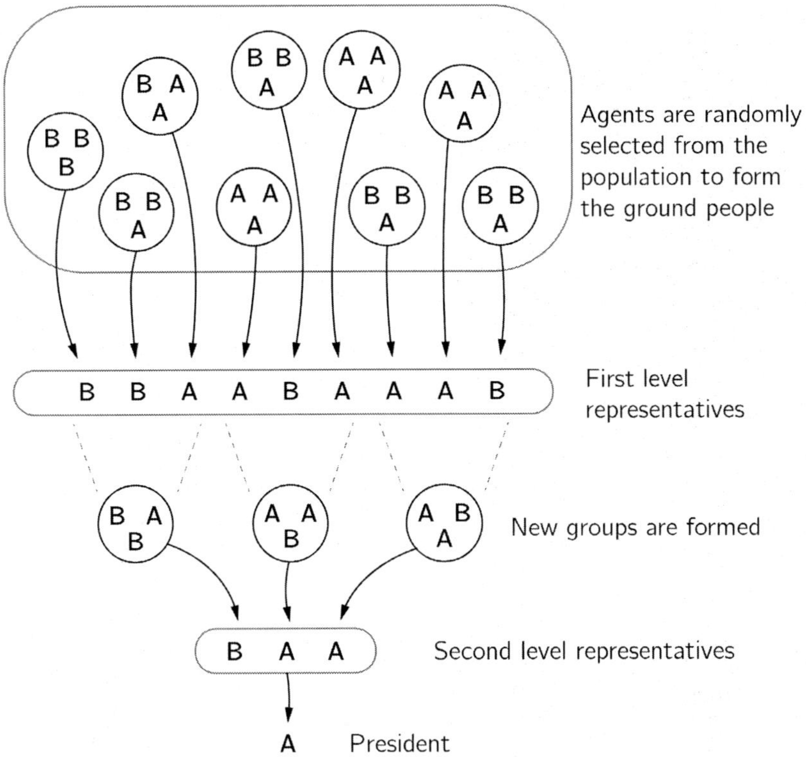

Figure 7.5. Bottom-up voting in hierarchical systems in the Galam model.

neighbors. The temperature variable in the Ising model is translated into randomness in the model. The external field is now interpreted as an external social field. In this way, one can explain phase transitions and hysteresis in opinion dynamics.

Another deeply analyzed option is the majority model (Galam 2008; Redner 2019). Here, a random group of voters is selected, and all voters in this group adopt the local majority opinion. This process can be repeated until convergence to one opinion is reached (which will always happen in a finite population). Galam (2008) sets up this process in a hierarchical fashion (see fig. 7.5). Alternatively, only one voter could be influenced by the majority vote in its neighborhood. This corresponds to the Ising model with 0 temperature.

Another case is the q-voter model (Castellano, Muñoz, and Pastor-Satorras 2009; Jędrzejewski and Sznajd-Weron 2019). Here, agents change their opinions only if all q voters selected from the neighborhood agree on the other opinion. When $q = 1$, this reduces to the standard voter model. The q-voter model generally allows for a higher degree of opinion diversity compared to the basic voter model. The q-voter model has been implemented in NetLogo ("qvoter_WS" in the user community models).

In the basic Snzadj model, agents are placed on a line, and two neighbors with the same opinion spread this opinion to their own neighbors. If they disagree, they enforce their disagreement on their neighbor. Thus $(?, 1, 1, ?)$ becomes $(1, 1, 1, 1)$ and $(?, -1, 1, ?)$ becomes $(1, -1, 1, -1)$. This can converge to a state of all 1s, all -1s, or a sequence of 1 and -1 pairs. The latter state is reached with a probability of .5. This model has also been extended in many ways, such as adding an election process (Sznajd-Weron, Sznajd, and Weron 2021).

7.3.1.3 Social Impact Theory

The last discrete model I mention here is the social impact model, which is based on Bibb Latané's (1981) psychological theory of social impact. Latané introduced many ideas and concepts from complex-systems theory into social psychology. His psychological theory is firmly grounded in social psychology and supported by all kinds of evidence (Karau and Williams 1993).

In this theory, opinion change depends on social impact I. Opinion X is either -1 or 1. Social impact is a function of the persuasiveness (p_i) of opponents (connected agents with the opposite opinion), the supportiveness (s_i) of supporters (with the same opinion), and the distance (d_{ij}) to these agents. The effect of distance can be modified with α. As the value of α increases, the influence of agents located farther away diminishes. All of these parameters are positive random values. The impact I is defined as:

$$I_i = I_i^P - I_i^S = \left[\sum_{j=1}^{N} \frac{p_j}{d_{ij}^{\alpha}} (1 - X_i X_j)\right] - \left[\sum_{j=1}^{N} \frac{s_j}{d_{ij}^{\alpha}} (1 + X_i X_j)\right]. \tag{7.2}$$

With j we take the sum over the neighbors of agent i. Note that when $X_i = X_j$, $I^P = 0$ due to the $1 - X_i X_j$ term, and the same is true for I^S when $X_i \neq X_j$. The effects of persuasiveness and supportiveness are reduced as the distance between agents increases. Setting α to values greater than 1 reduces the effect of distant neighbors. In addition to these forces,

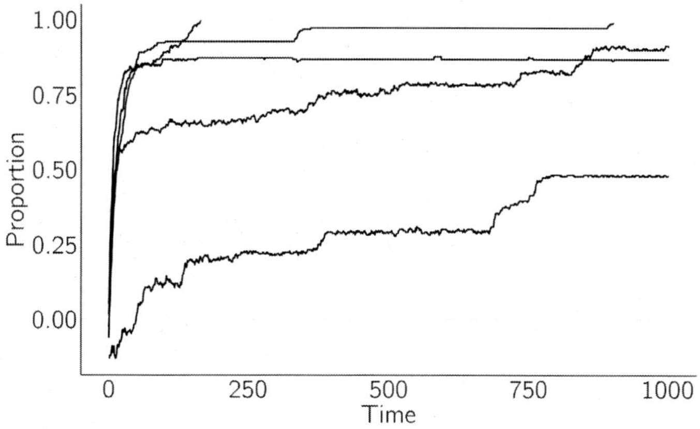

Figure 7.6. Five runs of the social impact model with $\alpha = 5$, p, s, and H sampled from uniform distributions between 0 and 100, and a lattice of 10 by 10. For example, the line at the top shows staircase-like behavior around time 320 and at the end of the time series. The Social Impact model can be found in the online NetLogo models and in the software repository of this book.

the theory assumes an external field H, as in the Ising model. The dynamic of opinion is:

$$X_i(t+1) = -sgn[X(t)I_i(t) + H]. \qquad (7.3)$$

Thus, opinion of agents become -1 if $X(t)I_i(t) + H$ is negative, and vice versa. Lewenstein, Nowak, and Latané (1992) present analytical mean-field solutions for fully connected networks. Without individual fields, the model ends up with an infinite number of stationary opinion states, one of which is usually dominant.

In the presence of individual fields, some minority opinions can become metastable. These smaller minority clusters can also persist for a long time before shrinking again, and the process repeats itself, resulting in what is called staircase behavior (fig. 7.6). Such a model can explain why small minority groups (such as flat-earth beliefs) often persist for a long time, against all odds (Douglas, Sutton, and Cichocka 2017).

Metastable opinions may persist for some time, but eventually, due to noise or other factors, they suddenly shrink to smaller clusters.

Extensions of this model include learning, leadership, external influences, and identity effects (for a review, see Holyst, Kacperski, and Schweitzer 2001).

7.3.2 CONTINUOUS OPINION MODELS

7.3.2.1 Classic Models

Another line of research, with its own history, starts from the assumption that opinions are continuous variables (for a review, see Noorazar 2020). They will have values between 0 and 1, for instance. A classical model is the DeGroot model, where agents are connected in a weighted network. At each iteration, an agent's opinion is set equal to the weighted average of all connected agents in the network. In this way, opinions tend to converge (figure 7.7). The Friedkin–Johnson model (Friedkin and Johnsen 1990) is an extension that includes a confidence level for each agent. This agent's confidence in their own opinion reduces the effect of others. Clustering or polarization in these linear models can only occur if parts of the network are unconnected. The Friedkin–Johnson model can be efficiently simulated with Grind (using the `method = 'euler'` option) by:

```
FJ <- function(t, state, parms){
  with(as.list(c(state, parms)),{
    X <- state[1:n]
    M <- M / apply(M, 1, sum) # weights sum to 1
    dX <- (1 - g) * M %*% X  + g * X - X
    return(list(dX))
  })
}
n <- 100
M <- matrix(runif(n^2, 0, 1), n, n)
g <- .95 # if g  = 0 => DeGroot model
x0 <- runif(n, 0, 1)
s <- x0; p <- c()
run(odes = FJ, method = 'euler', tmax = 100)
```

With an additional bias mechanism, in which confirming evidence is weighted more heavily relative to disconfirming evidence, polarization can also occur in connected networks (Dandekar, Goel, and Lee 2013).

7.3.2.2 Bounded Confidence

The bounded confidence mechanism has been extensively studied as the most effective way to generate divergence of opinions in continuous opinion models. It assumes that individuals have a limited willingness to accept

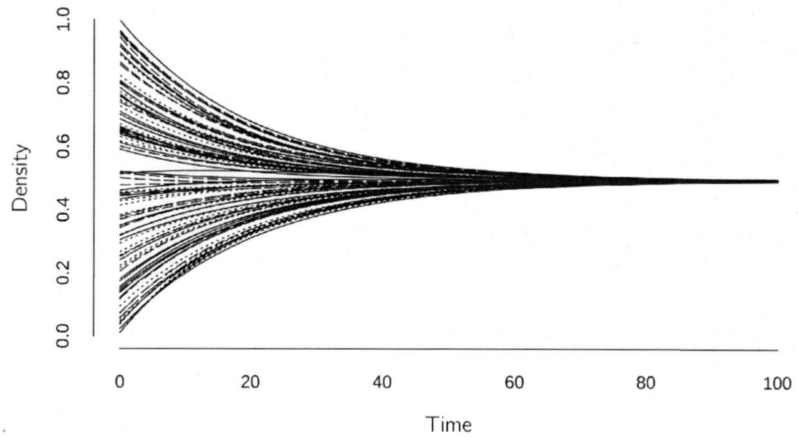

Figure 7.7. Convergence of opinion in the Friedkin–Johnson model in a connected network of agents.

and consider opinions that differ from their own and will only update their opinions if they are within a certain range or "bound" of similarity.

A simple but very interesting model is the Deffuant model (Deffuant et al. 2000). The initial opinions of n agents are randomly set to values between 0 and 1. At each step, two agents i and j meet. If $\lceil X_i(t) - X_j(t) \rceil > \epsilon$ nothing happens because the difference in opinion exceeds the bound ϵ. Otherwise, they exchange opinions according to:

$$\begin{aligned} X_i(t+1) &= X_i(t) + \mu(X_j(t) - X_i(t)), \\ X_j(t+1) &= X_j(t) + \mu(X_i(t) - X_j(t). \end{aligned} \tag{7.4}$$

So, if $\mu = .5$, they find each other in the middle. If $\mu = 1$, they take each other's position, as in the voter model. The value of μ does not make much difference, but the model converges fastest with $\mu = .5$. However, the choice of the bound ϵ makes a big difference. For $\epsilon = 0$, all agents stick to their positions; for $\epsilon > .5$, they all converge to $X = .5$. For intermediate values, different forms of clustering occur (fig. 7.8). It has been shown that the topology of the network does not make much difference (Fortunato

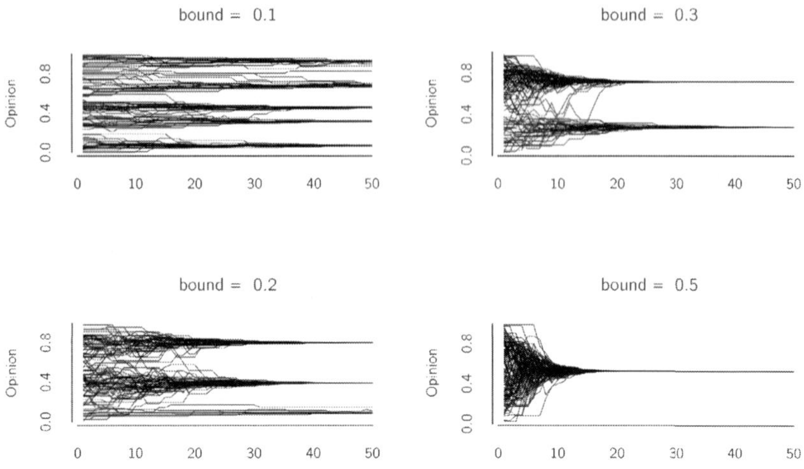

Figure 7.8. Four example runs of the Deffuant model with four different boundaries. Lower bounds of confidence lead to polarization.

2004). A drawback of this model is that it converges slowly. A fast but not entirely accurate code to simulate this model is: [1]

```
set.seed(20)
layout(matrix(1:4, 2, 2))
iter <- 50; mu <- .5; n <- 200
for (bound in c(.1, .2, .3, .5)){
  x <- runif(n, 0, 1)
  dat <- matrix(0, iter, n)
  for (i in 1:iter){
    # find an partner for every agent:
    y <- sample(x, n, replace = TRUE)
    x <- ifelse(abs(x - y) < bound, x + mu * (y - x), x)
    dat[i, ] = x
  }
matplot(dat, type = 'l', col = 1, lty = 1, bty = 'n', xlab = '',
        ylab = 'opinion', main = paste('bound = ',bound))
}
```

It turns out that adding some open-minded agents helps prevent polarization.

With this code you can explore many scenarios and variants. One interesting option is to have agents with different boundaries (Weisbuch et al. 2002). Also, adding some noise to X at each time step reduces polarization

[1] The second of the two equations is not implemented. This does not lead to different results as far as I know.

(Zhang and Zhao 2018). One can also lower the bound with the number of interactions. This increases the polarization (Weisbuch et al. 2002). One case I find interesting is increasing the bound after polarization emerged for a low bound. This gives hysteresis. A bound of .5 is sometimes insufficient to reduce polarization. Castellano, Fortunato, and Loreto (2009) review some other extensions (the role of propaganda, for instance).

Another well-known model is the Hegselmann–Krause model (Rainer and Krause 2002). This model is very similar to the Deffuant model, but instead of communicating with one other agent, they communicate with all connected agents, but only if the difference in opinion with these agents is sufficiently small. Thus, agents average the opinion of all connected agents for which the difference in opinion is less than the bound. This model is an extension of the DeGroot model and can be simulated by adding two lines to the Friedkin–Johnson code,

```
accepted <- abs(outer(X, X, '-')) < bound # acceptable neighbors
M <- accepted * M
```

after the `X <- state[1:n]` line and adding bound = .1 (fig. 7.9).

Again, many extensions have been studied. A recent paper studies the case where the network topology is a function of cognitive dissonance in opinions (Li et al. 2020). Baumann et al. (2020) present a continuous opinion model of echo chambers. Of particular interest is the multidimensional case (J. Lorenz 2007). When agents accept interaction based on the minimum distance along one dimension, consensus can be reached more easily. The idea of bounded confidence has been associated with the concept of the latitude of acceptance as proposed in social judgment theory (Sherif and Hovland 1961). This theory also proposes a latitude of rejection. There are two bounds, a lower and an upper bound. Below the lower bound, agents reduce their differences; between the bounds they ignore each other; and if they differ more than the upper bound, they increase their differences. This scenario has been investigated in Jager and Amblard (2005).

In the online library of NetLogo you can find the model "BC," which simulates both the standard Deffuant and the Hegselmann–Krause model.

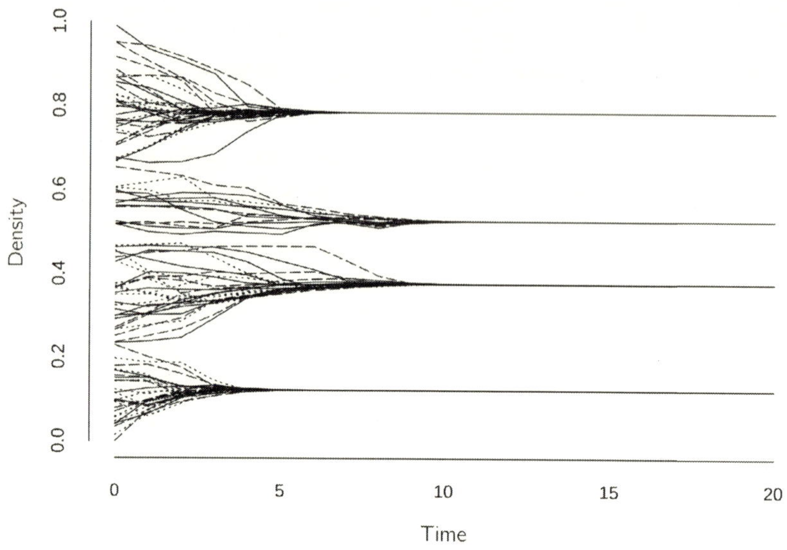

Figure 7.9. Clustering in the Hegselmann–Krause model.

7.3.3 EMPIRICAL VERIFICATION

Castellano, Fortunato, and Loreto (2009) note a striking imbalance between empirical evidence and theoretical models, in favor of the latter. It is not that there are no empirical data on the dynamics of opinions. Data come from studies on voting behavior, multicountry panel surveys, social media, and laboratory studies (for a review, see Peralta, Kertész, and Iñiguez 2022). The problem seems to be that these data do not discriminate between models. Most of the data fit all opinion models, supporting the general modeling approach but not specific models. This relates to the point that current opinion models are difficult to falsify because they lack specificity and are too flexibility. Flache et al. (2017) argue that the field suffers from a lack of systematic comparison of competing models. The theory-construction approach outlined by Borsboom et al. (2021) may be helpful here. We need a list of generally agreed-upon phenomena that all models are supposed to explain.

But another perspective is worth considering. In physics, models can be distinguished by their ability to make precise quantitative predictions. By contrast, in other fields there are often multiple models that roughly explain the same phenomenon, which can be advantageous. For example, Schelling's segregation model and its various iterations consistently predict that segregation occurs even when individuals are tolerant of different groups, demonstrating a robust prediction. Similarly, different models of traffic congestion tend to predict the same key phenomena. In addition, opinion models consistently show that zealots increase polarization, regardless of the specific model used. This convergence among different models may be the most reliable form of prediction we can achieve in these areas.

7.4 Psychosocial Models

In the introduction to this chapter, I said that psychology is the victim of the simplifications necessary to develop sociophysics models. In this section we explore ways to make these models a bit more psychologically realistic. Several existing models already include additional psychological variables (Jager 2017). The social impact model is a good example since it incorporates persuasiveness and supportiveness of agents to determine opinion change. The model is also based on a well-known theory in social psychology. Other models include stubbornness, cognitive dissonance, and confidence (Castellano, Fortunato, and Loreto 2009). All of these parameters are used to modify the interactions between agents.

Here I will discuss the model proposed in van der Maas, Dalege, and Waldorp (2020), which uses three dynamic variables to describe agents: information, involvement, and opinion. Individual agents are described by the cusp catastrophe as shown in figure 3.13. This is somewhat similar to the work of Sobkowicz (2012), who used the cusp model for the individual agent, using emotion and information as dynamic control variables. As in our model, interactions between agents change both opinions and the control variables. For example, agitation spreads across agents. However, in his opinion model, Sobkowicz reduces the cusp dynamics to a three-state system, where opinions are either -1, 0, or 1. We will not adopt this simplification, as much of the interesting dynamics (hysteresis within agents) are lost.

7.4.1 NETWORKS OF ATTITUDE NETWORKS

Our starting point is the Ising attitude model explained in chapter 6, section 6.3. In this model, attitudes or opinions are conceptualized as networks of feelings, behaviors, and beliefs about an issue. This new view of attitudes has been well received in the literature and applied to a number of attitudes (e.g., Chambon et al. 2022; Turner-Zwinkels and Brandt 2022; Zwicker et al. 2020). The idea is to use this attitude network model as a model for individual agents.

The resulting model becomes very complex—it is a network of networks model. This is not a new idea. Hierarchical or multilayer network models, such as multilayer neural networks (Treur 2019) and multilayer voter models (Masuda 2014), have been applied in many domains (for a review, see Boccaletti, Bianconi, Criado, del Genio, et al. 2014). However, such a model contains an enormous number of parameters and is hard to study. We take a simpler approach.

As discussed in section 6.3, the average behavior of the spins in the Ising model (the mean field) can be represented as a cusp catastrophe. This reduces the complexity enormously since the cusp attitude model contains only one equation with three variables: opinion (magnetization), information (external field), and attention or involvement (inverse of the temperature). Our model, the hierarchical Ising opinion model (HIOM), is an Ising-type social network in which each agent is a cusp. Interactions affect information and attention, leading to changes in opinion. We saw networks of interacting cusps in section 4.3. The HIOM is an extended form of this model.

The HIOM model can be found in the online NetLogo models (HIOM.nlogo) and in the online software repository of this book. In NetLogo, the equation and algorithm differ slightly from the original paper, mainly to speed up the simulation. The equations here are those used in the NetLogo model.

7.4.1.1 The HIOM

Like other opinion models, the HIOM makes assumptions about (a) the topology of the network, (b) the interactions between agents, and (c) the definition of opinion.

The qualitative results of the HIOM do not depend on the topology of the social network. In van der Maas, Dalege, and Waldorp (2020), the results are replicated for different topologies (e.g., 2D lattices, stochastic block models). However, as in other opinion models, more subtle results (e.g., convergence speed) are likely to depend on the network topology. For (b), the interactions between agents, specific assumptions are made, which are explained in the next section.

Regarding (c), the definition of opinion, the HIOM is special. Opinion is defined as a cusp. In the review of opinion models, I distinguished between discrete and continuous opinion models. Interestingly, the cusp behaves continuously for low values and discretely for high values of the splitting variable.[2] In this way, the HIOM bridges these two modeling traditions.

Whether opinion behaves discretely or continuously depends on another continuous variable (attention).

It is important to realize that the HIOM inherits assumptions from the Ising attitude model. First, the HIOM assumes that attitude nodes (representing feelings, beliefs, and behaviors toward the attitude object) are binary $(-1, 1)$. This is clearly debatable. Nodes might be better defined as $(0, 1)$ nodes, $(-1, 0, 1)$ nodes, or even continuous value nodes as in the XY model (Kosterlitz 1974). Second, we assume undirected pairwise interactions between nodes, whereas there is much to be said for directed effects. Third, attitude networks should be reasonably balanced (see sec. 6.3). To some extent, these assumptions can be relaxed without breaking the link to the cusp (see the appendix of van der Maas, Dalege, and Waldorp 2020).

In the HIOM, information and attention are updated based on interactions between agents. Information and attention are two orthogonal axes in the cusp. Information summarizes all variables and influences operating along the normal axis of the cusp. Its neutral value is 0, and negative and positive values are associated with negative and positive opinions. Attention has nonnegative real values. The opinion of agent i at time t changes according to the cusp equation with information and attention as control variables. For the implementation in NetLogo, I write the equation as a stochastic difference equation:[3]

$$O_{i,t+1} = O_{i,t} - \left(O_{i,t}^3 - (A_{i,t} + A^{min})O_{i,t} - I_{i,t}\right) t_s + \epsilon_{i,t+1}. \quad (7.5)$$

[2] This is highly relevant to the discussion in psychology about type and continua, that is, whether psychological traits are typological or continuous constructs (Borsboom, et al., 2016). They can be both!

[3] To incorporate close to linear change in O as a function of I, I use $A + A^{\min}$, where $A^{\min} = -.5$ and $A \geq 0$. See the original paper for explanation.

The ϵ term represents white noise sampled from a normal distribution $N(0, s_O)$. The time step, t_s, in this equation is set to a low value (.01) to prevent oscillations and chaotic behavior.

7.4.1.2 Agent Interactions: Information and Attention

The HIOM makes three assumptions about interactions. First, it assumes that agents initiate interactions based on their involvement. The idea is simply that I'm not likely to start a discussion about a topic—say, about genetically modified food—if I'm not interested in the topic. The probability of initiating an interaction is equal to attention:

$$P_{i,t}(initiates\ interaction) = A_{i,t}. \tag{7.6}$$

Second, it assumes that the attention or involvement slowly decreases over time. Attention or involvement is a limited resource; one cannot be involved in everything all the time. With the constant emergence of new interests or topics, attention to older topics tends to wane.

Third, attention increases again through social interactions. If someone starts a conversation about genetically modified food, my interest in the topic is likely to increase. A simple way to implement this is:

$$A_{i,t+1} = decay_A(A_{i,t} + d_A u_{i,t}). \tag{7.7}$$

When the agent is involved in an interaction, we set $u_i = 1$; otherwise $u_i = 0$. The parameter d_A determines the rate of change of A due to interactions. The decay in attention, $decay_A$, is applied to all agents.

The fourth assumption is about information. We assume that the exchange of information is an averaging process weighted by attention. If agent i is less attentive to the attitude object than agent j, agent i will move more to the information position of j than j will move to i.

This is formalized by:

$$I_{i,t+1} = r_t I_{i,t} + (1 - r_t) I_{j,t},$$
$$\text{where } r_t = r_{\min} + \frac{1 - r_{\min}}{1 + e^{-p\left(A_{i,t} - A_{j,t}\right)}}. \tag{7.8}$$

I_i and I_j denote the information of agents i and j, the agents involved in the interaction. Resistance, r in [0, 1], determines the relative impact of agent j on agent i. Resistance or stubbornness is a logistic function of the difference in attention between the agents. Thus, if $A_i \ll A_j$, r will be close to 0 and the information in agent i will change to the value of the information in agent j. The strength of this effect, persuasion, is determined by the steepness, p, of the logistic function. The parameter $r_{\min}$ determines the minimal value of r. If $r_{\min}$ is high, r will be high and agents will stick to their information state.

In some scenarios it is of interest to allow for decay in information especially in combination with a normally distributed, $N(m_I, S_I)$, noise term, ϵ. The full equation for the update of information is:

$$I_{i,t+1} = decay_I\left((1 - u_{i,t})I_{i,t} + u_{i,t}(r_t I_{i,t} + (1 - r_t)I_{j,t}) + \epsilon_{i,t+1}\right), \tag{7.9}$$

$$\text{where } r_t = r_{\min} + \frac{1 - r_{\min}}{1 + e^{-p\left(A_{i,t} - A_{j,t}\right)}}.$$

If $m_I \neq 0$ and $s_I = 0$, a constant external field is active. If $decay_I < 1$, information gradually shrinks to 0.

7.4.1.3 Algorithm

The model is now in place and can be simulated by following the steps below:

- Select a network topology.
- Set the model parameters t_s (.01), $A_{\min}$ ($-.5$), s_O (.01), m_i (0), s_i, d_A, p, and $r_{\min}$. Values in parentheses are defaults.
- Initialize agents, set I_{init}, A_{init}, and O_{init}.
- Iterate:
  - Randomly choose a set of agents, weighted by attention A (eq. 7.6).
  - Iterate over this "active" set of agents ($u_i = 1$}:
    - For each agent, randomly choose a neighbor as partner in the interaction.
    - Add attention to both agents (eq. 7.7, $+d_A u_i$).

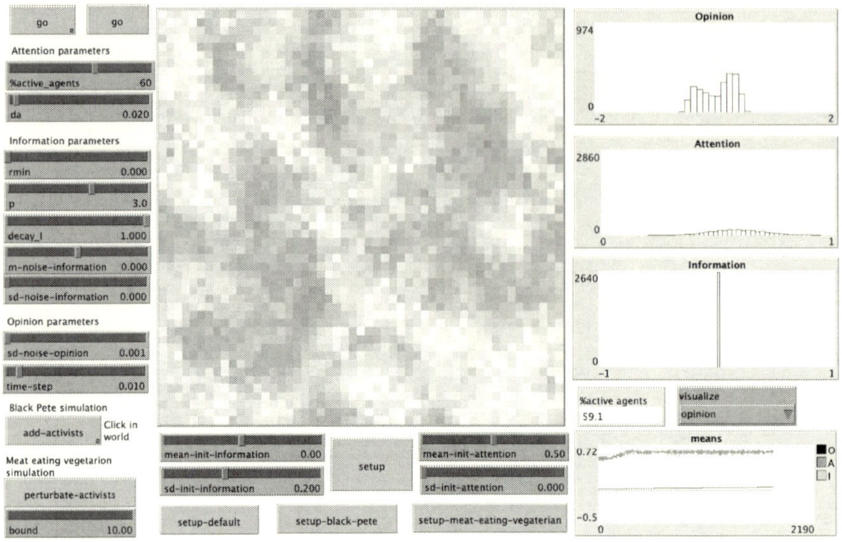

Figure 7.10. The interface of the HIOM NetLogo model. All model parameters can be adjusted with sliders. With **add-activists** and **perturbate-activists** , the key effects of HIOM can be reproduced. The graphs on the right show the distribution of opinion, attention, and information and the change in the means of these variables. Refer to the online version of the book for a color image.

  * Exchange information (equation 7.8).

- Apply decay in A to all agents (eq. 7.7, $decay_A A_i$).
- Add noise and apply decay in I to all agents (eq. 7.9, $decay_I$ $(I_i +$ $(0, s_I))$.
- Update opinion O in all agents (eq. 7.5).

The $decay_A$ has a special role in the current implementation of the HIOM. Instead of being fixed, it depends on the difference between the percentage of agents in the "active" set and the desired percentage of active agents (*%active_agents*), which is controlled by one of the sliders in the NetLogo model (fig. 7.10). This allows us to manipulate the general interest (attention) in the opinion object.

There is no obvious stop criterion, but in practice some type of convergence happens over time.

In the standard setup, you can see that high attention (set `%active-agent` high) leads to polarization. If you decrease the difference in information (set `decay_I` to .5), the polarization remains even if the difference in underlying information is 0. Only if you also decrease attention (set `%active-agents` low) does the polarization disappear. This is the first simulation described in van der Maas, Dalege, and Waldorp (2020). The second and third simulation are described in the next two sections.

7.4.1.4 The Persuasion Paradox

Suppose we have a network of conservative but low-attentive agents into which we add a few highly attentive activists with the opposite information and opinion. At first glance, one would expect the activists' opinion to spread quickly. They initiate all interactions (according to eq. 7.6) and have much more impact on their interaction partners than vice versa (eq. 7.8). However, this is not what happens. This scenario leads to polarization. The attention of some conservatives is too quickly raised, and people get a strong opinion against the activists. Figure 7.11 shows this effect which of course, depends on some parameter settings. If attention increases more slowly, activism may spread better. You can test this scenario in the NetLogo model by selecting `setup-black-pete` and `add-activists`. Activists can now be added by clicking in the world. Vary persuasion p and d_A (da) to see different effects.

What is new in this model? It shows a new mechanism for polarization, namely hysteresis within agents, leading to the persuasion paradox. Without a cusp for the individual opinion dynamics, this form of polarization would not occur.

The persuasion paradox describes how efforts to persuade can backfire, causing the other person to become even more entrenched in their original position. This is due to the effect of social interaction on attention.

7.4.1.5 A Counterintuitive Prediction

In continuous-opinion models, the main reason for polarization is bounded confidence. We can easily add this to the HIOM: when $|O_i - O_j| > \epsilon$, there is no interaction (no increase in attention and no exchange of information). This would even increase the polarization in the HIOM. However, since opinion and information can have opposite signs due to hysteresis, there is a way out. An example would be a meat-eating vegetarian, a person with a vegetarian point of view who actually eats meat. I claim to be such a vegetarian, but this position is not generally accepted in my local environment.

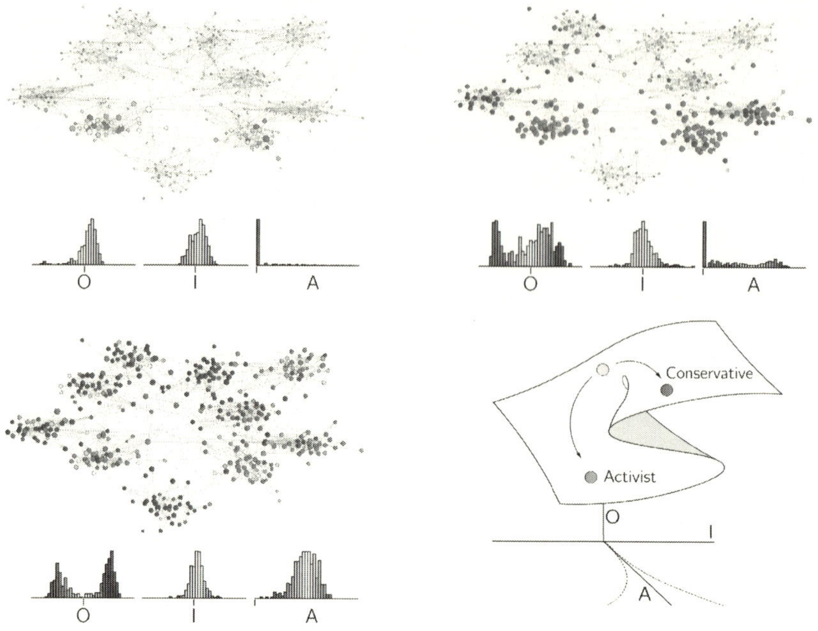

Figure 7.11. The visualization of persuasion paradox. Activism initially spreads quickly but also increases the attention of conservative agents (see second and third panel). Over time, opinion polarizes. Red nodes are activists and blue nodes are conservatives. The size of the nodes represents attention. The last panel explains why some conservatives become (anti-)activists themselves. Refer to the online version of the book for a color image. *(Reprinted from van der Maas, Dalege, and Waldorp (2020) with permission.)*

The scenario, simulated in van der Maas, Dalege, and Waldorp (2020), consists of a majority of low-involved meat eaters and highly involved vegetarians. The meat eaters refuse to talk to the vegetarians due to bounded confidence, leading to increased polarization.

Now some attentive vegetarians will be perturbed into the meat-eating position. This is a metastable state. This state is somewhat stable, although they may jump back after a few iterations. But while the vegetarians are in this metastable meat-eating state, they can communicate with the less involved meat-eaters and spread vegetarian information. This proves effective (see fig. 7.12). This scenario is also implemented in the NetLogo model.

7.4.1.6 Variants of the HIOM

What appeals to me about the HIOM is its ability to incorporate a greater degree of psychological insight into opinion dynamics models, resulting in

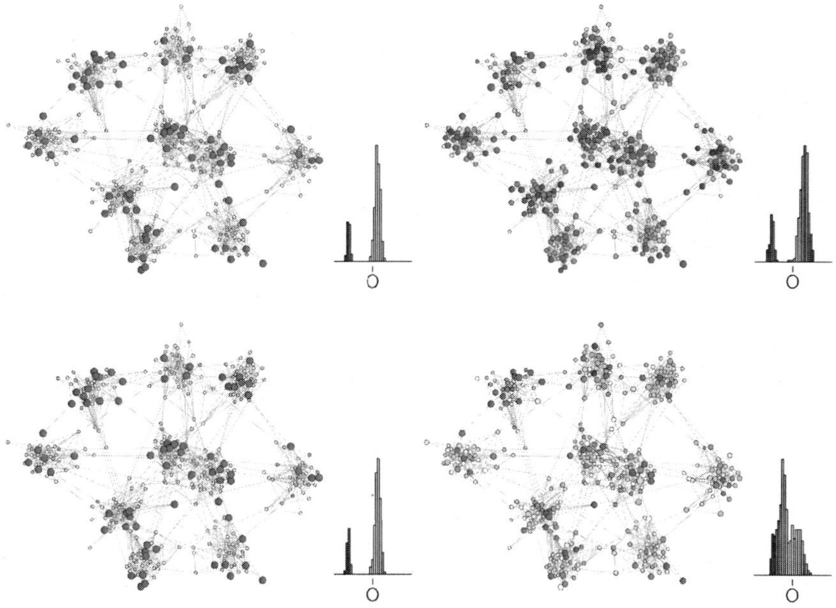

Figure 7.12. The meat-eating vegetarian. The left panels show equal initial states. In the top panels, bounded confidence prevents the vegetarian (red) and the meat-eater (blue) from interacting. This leads to polarization. In the bottom panels, some vegetarians are perturbed toward meat eating. These perturbed agents have $I < 0$ and $O > 0$, which is possible because $A > 0$ (hysteresis). Since $O > 0$, these agents can exchange information with meat-eaters, leading to the spread of vegetarianism. Refer to the online version of the book for a color image. *(Reprinted from van der Maas, Dalege, and Waldorp (2020) with permission.)*

a novel explanation for polarization attributed to hysteresis within individuals. A strength of the HIOM is that the definition of opinion is based on the Ising attitude model, a psychological network model of attitudes and opinions that has been developed and supported in a number of papers (see sec. 6.3). Furthermore, it presents an untested, counterintuitive prediction that warrants further investigation.

What is also unique about this model is the role of attention/involvement. It shows the risks of too-low and too-high values of this variable. Too-high values lead to extreme hysteresis effects in agents, making a change of opinion impossible. This is clearly a dilemma for activists. But too little involvement is also risky. The moment you stop influencing such agents, the attitude nodes start behaving randomly again, and they will move back to the neutral opinion at the back of the cusp. The per-

Talking to a person with very low involvement may have a positive short-term effect, but no long-term effect.

suasion paradox and the involvement dilemma need further study, but they seem important.

What I like less is that the HIOM is perhaps too complex, which makes it difficult to study its behavior. One idea is to make attention a network parameter. An attentive agent simply has many connections to other agents. As attention decreases, this translates into fewer connections.

A more radical simplification that still uses the attention/involvement idea is a voter model with three votes (leftist, centrist, and rightist), as in the constrained three-state voter model (Redner 2019). Leftists and rightists don't talk to each other (bounded confidence), but they do talk to the centrists. What we can add to this model is attention. Extremists (left and right) have high attention, while centrists have low attention. As in the HIOM, more attentive agents are more persuasive, so centrists tend to become more extreme due to interactions. On the other hand, attention is costly, and there is a probability that extremists spontaneously decay to the centrist position. Such a model can be studied analytically. This voter-type model has no within-subject hysteresis and will show less interesting behavior.

What I also like less is that the HIOM is not complex enough. A huge simplification is that we have left out the fact that we have more than one attitude. We have hundreds or maybe thousands of attitudes. Since these are not independent, it is safe to say that these attitudes form complex networks with subclusters and central (hub) attitudes. A better model would be a (social) network of (attitude) networks of (attitude element) networks. In our current work, the level of within-person attitude networks is underexplored. Finally, many crucial aspects of opinion formation are missing, such as the role of (social) media, confirmation bias, identification with groups in society, the political system, etc.

What I really don't like is that the Ising attitude model, in which attitudes behave according to the cusp, always leads to radicalization when we get involved. At best, we can jump irregularly between the extreme states (as in fig. 4.2) as a form of ambivalence. But our society needs people, for example, judges, who get involved and attend, but at the same time remain neutral. This is impossible in the Ising attitude model and thus in the HIOM.

A possible way out is to start from the tricritical Ising or Blume–Capel model (see sec. 6.3), where the attitude nodes have three states $(-1, 0, 1)$ instead of only two. The resulting dynamic equation is the butterfly catastrophe, which has four instead of two control variables. Building this in the HIOM will be a challenge. The Blume–Capel model has also been proposed as a between-person opinion model (Barbaro, Chayes, and D'Orsogna 2013; Ferri, Díaz-Guilera, and Palassini 2022).

7.4.2 CASCADING TRANSITIONS IN OTHER PSYCHOSOCIAL SYSTEMS

We have now seen two psychological models of cascading transitions. The first was the model for multifigure multistable perception (sec. 4.3). This was a within-person model. The second is the HIOM. But many other processes in psychosocial systems come to mind. One example is addiction. At the psychological level, the dynamics of addiction are often sudden (e.g., quitting and relapse), while at the social level there are sudden outbreaks of substance abuse (e.g., the heroin epidemic).

To model this cascading process, we could follow the same approach as in the HIOM. Instead of the cusp, we could use the spruce budworm model as a model of individual substance use. Interactions in the network affect the parameters r_b, K, A, and B parameters (Boot et al., submitted for publication). A similar approach is possible with the panic model (sec. 4.3). In future work, we will also apply this model to collective learning processes.

7.5 Psychosociophysics

Neglecting the social world while attempting to model complex psychological processes may lead to inaccurate outcomes (Sobkowicz 2020). It is crucial for psychologists engaged in modeling to familiarize themselves with prominent sociophysics modeling techniques. The objective of this chapter was to offer a comprehensive overview of these models, but it is important to note that this review is not exhaustive. As new models continue to emerge, the foundational knowledge provided here will, I hope, enable readers to stay informed and assess their applicability to psychological inquiries.

Similarly, excluding psychological factors while modeling the social world can be counterproductive. It is essential to emphasize the need for psychosocial or even psychosociophysical models that integrate both aspects. The social impact model is as an example of such an approach. Our own model, the HIOM, has been elaborated upon in this chapter. The HIOM stands as a significant step in the direction of psychosociophysics models. The concept of cascading transitions is evidently relevant to other applicable cases, and there is ample opportunity for further exploration in this field.

7.6 Exercises

1. Rerun the Schelling simulation (figure 7.1) with density of 50 instead of 97. Are the results substantially different? Submit your plot. (*)

2. Implement the equations (equation 7.1) for the simple naming game in Grind. Show that the case where all three options A, B, and AB coexist is an unstable fixed point. (**)

3. What happens in the Axelrod model (sec. 7.2) when $F = 10$ and $Q = 1$. Why? (*)

4. Implement the 1d voter model (section 7.3) in either R or NetLogo (using BehaviorSpace and R for data analysis). Check that the probability of convergence to a state is equal to its initial proportion. Then check that the convergence time is a quadratic function of N by plotting the square root of this time versus N. Take the average of these times for at least twenty runs. (**)

5. In the social impact model (sec. 7.3), does the staircase behavior depend on the max-h parameter? (*)

6. In the Deffuant model (section 7.3), a limit of .5 almost always leads to convergence of opinions. Adjust the R code for the Deffuant model so that the bound grows from 0 to .5 over 1,000 iterations. You end up with one or two clusters. Why do you end up with two clusters even if you increase the number of iterations? (**)

7. Run the HIOM NetLogo model (sec. 7.4). Set the `mean-init-information` to .5, the `mean-init-attention` to 1, the

`bound` to .2, and the `%active-agents` to 90. Let it run and use `add-activists` and `pertubate activists`. Why does this not result in a change? (*)

8. Think of a simple way to add the effect of media to the HIOM. Implement and present your results. (**)
9. Implement the HIOM in a preferential attachment network. Use the "Preferential Attachment NetLogo" model. (**)
10. Design an empirical study to test the "meat-eating vegetarian" prediction (sec. 7.4).

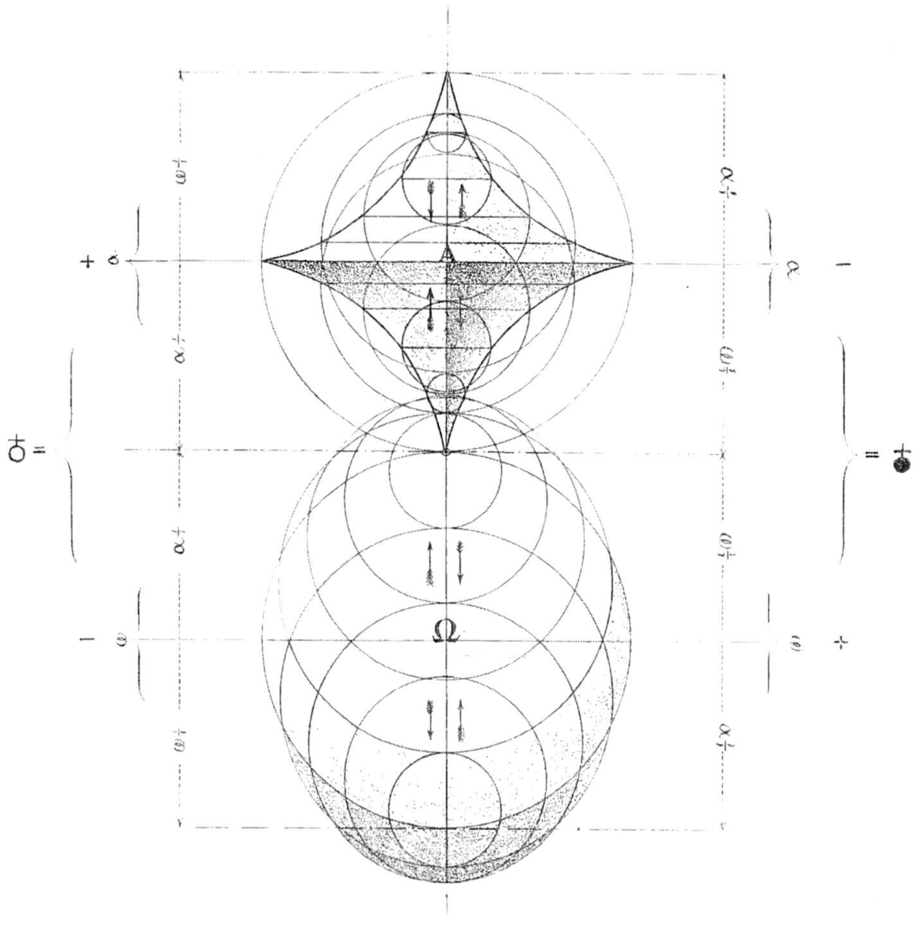
Ω

6

EPILOGUE

In writing this book, I had three primary goals: to provide a comprehensive overview of complex-systems research with a particular emphasis on its applications in psychology and the social sciences, to teach skills for complex systems research, and to encourage critical thinking about the potential applications of complex systems in psychology. I will first discuss the main points of the chapters briefly and then focus on evaluating the complex-systems approach to psychology.

RESOURCES
The electronic version of this book, including code blocks and links, can be viewed and accessed at: *santafeinstitute.github.io/ComplexPsych/*

In chapter 1 I defined complex systems. Complex systems are composed of interacting subsystems, resulting in emergent behaviors not seen at lower levels. These emergent patterns often arise through self-organization, can change suddenly, and may exhibit chaotic behavior, making prediction difficult. They can be studied using various methods, such as those originating in nonlinear dynamical system theory, agent-based modeling, and network theory.

I argued for the independence and autonomy of psychology as a scientific discipline. This does not mean that the mechanisms and principles that operate in psychological systems are necessarily different from those in other sciences. The synchronization of atoms in a laser beam may not be very different from the alignment of symptoms in a depressed patient or the synchronized movements of people evacuating a burning building. Complex systems in any science are almost always networks in which unpredictable phenomena can occur. These can be networks of atoms, neurons, symptoms, cars, people, or even entire countries. In these dynamic networks, there are usually only a limited number of organized patterns or stable states. Manipulating control variables can lead to bifurcations, which can take a limited number of forms. We have seen these principles again and again.

The power of the complex-systems approach is that the same organizing principles can operate at very different levels of description.

One may argue that emergent properties at one level of description can always be explained or rewritten as lower-level events, but this is largely irrelevant. In psychology we have to distinguish at least three basic levels of description related to the brain, the mind, and the social world. In describing the mind, delving deeper than the level of neurons does not provide additional insights.[1] And in modeling the spread of conspiracy theories, the way axons of neurons grow can be left out of the equation.

I ended with reasons to be moderately optimistic about the application of complex-systems science in psychology. First, complex systems can often be simplified without losing their explanatory value. Second, these systems can be described by a limited number of equilibria. Third, we can use network science to model complex systems.

In chapter 2, I introduced chaos theory and many of the key concepts of complex systems, such as fixed points, limit cycles, bifurcations, and dynamical system models. Learning about deterministic chaos changed my worldview and also my appreciation of beauty in mathematics. I hope I succeeded in conveying both.

The principal unpredictability of complex systems, when they are in the chaotic phase, is important to understand.

It often seems to me that psychologists somehow believe that if they could collect vast amounts of extremely accurate life history, environmental, genetic, biological, and neuropsychological time-series data from millions of individuals—a feat that is currently unattainable—and feed it into a sophisticated nonlinear multilevel regression model with numerous higher-order interactions, they could predict virtually anything. This is simply wrong, not only because of deterministic chaos but also because of the influence of epigenetic processes (the third source), which I discussed in section 4.3. The Pólya urn model provides a very simple demonstration of why even perfect knowledge of the initial conditions and dynamics of a system is insufficient to predict individual developmental outcomes.

[1] Actually, the role of quantum processes in psychological processes is a topic of ongoing debate and research in the scientific community. While some researchers have proposed that quantum mechanics might play a role in cognitive processes, this idea is not widely accepted. One theory suggests that quantum processes in microtubules within brain cells could be linked to consciousness (Hameroff and Penrose 1996). Other researchers have suggested that the probabilistic nature of quantum mechanics might provide a better model for human decision-making than classical probability theory (Busemeyer and Bruza 2012). However, this work does not suggest that quantum physical processes are involved in decision-making. Rather, it uses quantum theory as a mathematical framework for modeling cognitive processes.

The application of chaos theory in psychology is limited to the analysis of psychophysiological data. The hypothesis is that the brain works best on the edge of chaos (see also sec. 5.4). Despite hundreds of papers written in the last forty years, I would say that the evidence for this hypothesis is inconclusive. This approach requires very high-quality data and advanced statistical approaches, which are difficult to acquire and challenging to develop. Furthermore, it is worth mentioning that there are even skeptics who question the applicability of chaos theory in studying complexity altogether (Anderson 1999a).

Chapter 3 was devoted to transitions, or tipping points, as they are sometimes called. To me this is a very practical concept in complex-systems theory. I am probably somewhat biased, but I see instances of tipping points across a wide range of psychological processes. I list just a few that I have not discussed: sudden insights, creative breakthroughs, aggressive acts, the onset of puberty, mood swings, vocabulary spurts, and dropping out of school. In all these cases, the modeling and empirical program laid out in chapter 3 might be fruitful. The chapter also contained an elementary introduction to the mathematics of bifurcations and catastrophes. I think a basic understanding of the key formal concepts is necessary and achievable. I provided sources for further study; for those who find these concepts still difficult, I also recommend running and studying the examples of the cusp catastrophe in chapter 4. I have also outlined a methodology for empirically evaluating catastrophe models using catastrophe flags and Cobb's statistical approach.

I believe that this methodology effectively addresses the major criticisms of the application of catastrophe theory in the social sciences.

In chapter 4, I focused on building dynamical systems models. I introduced Grind as a tool for implementing a wide range of dynamical systems models in biology and psychology. This modeling approach allows us to build more mechanistic models that we can still fully understand (using numerical bifurcation analysis). I have tried to give a representative overview of dynamical systems modeling in psychology, but you will easily find many other models in the literature. In such a case, I always recommend implementing the model yourself. In most cases, Grind will do. Replication is the key to good science, and you will learn a lot in the process.

I discussed the evaluation of ecosystem models in some detail (sec. 4.2). I noted that even simple models imply a large number of assumptions, many of which are made implicitly (with the assumptions underlying the Lotka–Voltera model as an example). Also, seemingly trivial changes in model choices can have a huge impact on the qualitative behavior of models. As I said in the same chapter, I find the process of formalizing a verbal model fascinating. It tends to be very confusing. Suddenly it is unclear what the assumptions are, what mechanism is really being proposed, what the time scales are, or even what the phenomenon to be explained really is.

It is essential to have studied many different examples of dynamical systems models before you attempt to build your own.

Models usually combine a number of mechanisms, and I strongly recommended reusing mathematical model pieces in other models. Modeling by analogy can be very productive (Haig 2005). Finally, when modeling, it is critical to always keep the connection to the data in mind. I am particularly concerned about this with the causal loop diagram approach. Models built in a session with content experts tend to get big with lots of boxes and links. This may not be a problem if all time series of measurements for each of these boxes are available, but this is rarely the case. I prefer to start simple and only add variables and equations when some established phenomena cannot be explained by the most trivial model.

In chapters 2, 3, and 4, I focused on systems with a small number of variables. This part of the book covers what is often called nonlinear dynamical system theory. I view the theory of nonlinear dynamical systems as a fundamental component of the complex-systems approach. The second part of the book deals with systems with a large number of variables.

Self-organization was the subject of chapter 5. I also used this chapter to introduce main theories in the study of complex systems, such as Haken's work on synergetics and Prigogine's ideas on irreversible transition and the second law of thermodynamics. I hope that I have successfully conveyed my own sense of awe and amazement at the self-organizing processes found in nature, as well as the unexpected potency of seemingly simple systems like the Game of Life. By engaging in the NetLogo simulations, I trust that the somewhat abstract concept of self-organization has become more comprehensible. As an example, I mention the spiral waves in the spatial model of hypercycles that prevented the abundant growth of parasites. This example of strong emergence is easy to understand by running the simulation and studying the basic NetLogo code.

I again realized while writing this book how the concepts of self-organizing complex systems were already present in early day psychology. I gave Gestalt psychology and Piagetian theory as examples, but one could make the same point for Rogers's theory of self-concept, Heider's balance theory, or Gibson's ecological theory of perception. This is why I perceive the complex-systems approach in psychology less as a novel theory and more as a formalization of these intriguing yet abstract verbal theories. For instance, the Ising attitude model formalizes of numerous established concepts in social psychology.

The concepts of self-organizing complex systems were already present in the early days of psychology.

Chapter 6 focused on the psychological and psychometric network approaches. This chapter ends with an extended discussion of the challenges facing this popular research line. My own contributions are mainly theoretical. I do think it is important to provide an alternative to the common cause view on psychological traits. The inability to intervene on common causes, or even to gain knowledge of these fixed biological factors without relying on the observable factors they explain, leads to a discouraging psychological theory that can easily be misused to abandon the less fortunate to their fate (Heckman 1995).

In terms of modeling, I consider the application to attitudes to be the most successful. The Ising attitude model not only builds on earlier connectionist network models, but also incorporates improvements. The dynamics of the Ising model are better understood from a mathematical perspective, it offers a novel psychological interpretation of the temperature parameter, and it can be effectively fit to data. The model formalizes key concepts in social psychology, such as dissonance and the mere thought effect, while suggesting a new explanation for the differences between implicit and explicit attitude measures. Using the mean-field approximation of this model in the HIOM represents an innovative development in sociophysics. I expect many more innovations in both network psychology and network psychometrics.

Equating attention and (inverse) temperature may have broader implications beyond attitude theory.

In the final chapter, chapter 7, we moved into the realm of the social world. Because psychologists are often unfamiliar with disciplines such as computational social science, sociophysics, and agent-based modeling, I aimed to provide a concise overview of this rapidly evolving field. Finding the right balance in simplification has proven to be a challenge. While the simple voter model offers analytical tractability, its relevance is primarily theoretical. More realistic models quickly become intractable, even with the help of simulations.

I have focused on improving the psychological realism of the agents. My approach involves linking three levels of description: the interaction of attitude elements, the mean-field representation of attitudes as cusps, and the HIOM, where cusp-like agents interact within social networks. To my knowledge, this three-level integration is unique. Similar to many complex models in psychology and the social sciences, a notable weakness is the connection to data, specifically, data that really differentiate models.

I expect to see many more applications of the cascade transition model to psychological systems.

Let's zoom out more. By now, you should have a solid understanding of what complex systems involve, including the concepts of deterministic chaos, catastrophes, and self-organization. I have illustrated these ideas with numerous examples from various scientific fields, including psychology. It is vitally important to be aware of existing models and frameworks when you begin constructing your own models. This was my first goal.

Second, I placed significant emphasis on the importance of modeling skills. For me, the one critical path to comprehension lies in simulation. I hope that the skills you have developed by working through my exercises will encourage your engagement in formal modeling in psychology.

My third objective was to encourage a critical approach toward the study of complex systems in psychology and, in a larger context, within the realm of psychology and science itself. This naturally leads to the question of my personal stance on this undertaking. I must confess, I have mixed feelings.

About twenty years ago, in my inaugural lecture, I portrayed myself on the unstable maximum between the two minima of the cusp potential function. The attitude object was our field of psychology, and the minima represented positive ("love") and negative ("hate") evaluations. Since I am obviously highly involved in the matter, I'm on the front side of the cusp, which means that this in-between state is highly unstable, assuming, of course, that the cusp model of attitudes is correct.

So, if you are asking me to burn our field to the ground, including my own work, you have come to the right place. In short, we have a replication crisis (Aarts et al. 2015; Nosek et al. 2022; Pashler and Wagenmakers 2012), we have a theory crisis (Eronen and Bringmann 2021; Oberauer and Lewandowsky 2019), and we have a measurement crisis (Franz 2022; Lumsden 1976; Michell 1999). How many crises can you have?

Each of these crises is more severe than one might think at first glance. For a long time, questionable research practices, *p*-hacking, selective reporting, cherry-picking studies, presenting exploratory results as confirmatory results, to name a few, dominated research. All of this has changed radically since 2011. Surprisingly, a fraud case[2] in my own country played a major role in this shift to open science, preregistration, and data sharing. Maybe we were in a metastable state and just needed one such perturbation. We are still in the midst of this transformation and should not be celebrating too soon (Chambers 2017). But a revolution it is!

We can be most optimistic about the replication crisis.

Then the theory crisis. Depending on the criteria (weak or strong), psychologists either have millions of theories or none. There is not much in between. Many recent papers have proposed formalization as a way out of the theory crisis (Oberauer and Lewandowsky 2019; Borsboom et al. 2021; Rooij and Blokpoel 2020). This book is written from that perspective. I hope it makes a contribution, but I'm well aware of the differences between the scientific basis of formal models in the natural sciences and our modest attempts. To me, it is all about the right degree of simplification. In modeling complex systems, we almost always define at least two levels: the microscopic and the macroscopic. Emergent phenomena at the macroscopic level arise from microscopic interactions. The step from neural activity to higher reasoning just seems too large (although we may need to rethink this in light of the astonishing successes of large language models).

In writing this book, I learned that the distinction between phenomenological and mechanistic modeling is less discrete than I thought. First, we have been able to provide a foundation for some phenomenological catastrophe models. The cusp model of attitudes can be derived from the Ising model of attitudes, which is based on some simple assumptions about attitudinal networks. Second, many biological models combine phenomenological and mechanistic elements. Some terms may be well argued, others are just pragmatically chosen (the Holling types, for example). Nevertheless, I hope to have shown that studying these

[2] https://en.wikipedia.org/wiki/Diederik_Stapel

models is very informative. In my work, analogical modeling plays a central role.

I see the measurement crisis as the most serious problem. Let me recall Richard Feynman's claim that the accuracy of calculating the size of the magnetic moment of the electron is the equivalent of measuring the distance from Los Angeles to New York, a distance of over 3,000 miles, to within the width of a human hair. And that was in 1985! I am an expert in psychological measurement and have published many papers on new methods of psychological measurement. I can tell you that we are nowhere near this amazing level of quantification and precision. We cannot do addition!

Addition is the litmus test of quantification (Michell 1997). Real quantities, such as weights and distances, can be added. One kilogram + one kilogram = two kilograms, which is twice as much as one kilogram.[3] We cannot say such things about IQs, personality test scores, or Likert scores on attitude items. I do not think this is a hopeless endeavor. Even physicists have lived through times when key concepts were vaguely understood and poorly measured (see *Inventing Temperature* by Chang (2008)). I also draw hope from statements such as "Every law of physics, pushed to its extreme, will turn out to be statistical and approximate, not mathematically perfect and precise" (Wheeler 1994).

Currently, our scales of measurement are somewhere between ordinal and interval. Perhaps there is a continuous path to improvement. But for now, we have to live with rather weak scales of measurement. This has implications for our attempts to formalize psychological theories. The exact form of certain terms in our equations is irrelevant when we have only ordinal or semi-interval data. When testing models, we should focus on their qualitative behavior. This is exactly what we did with the catastrophe flags in chapter 3, and it is one reason why I adhere to the sometimes-criticized catastrophe theory.

It is also important to temper expectations about distinguishing models based solely on their quantitative predictions. Achieving consensus on

[3] This is sufficient but not a necessary condition. Sometimes a concatenation of two quantities gives a weighted mean, for example when blending two liquids with varying temperatures.

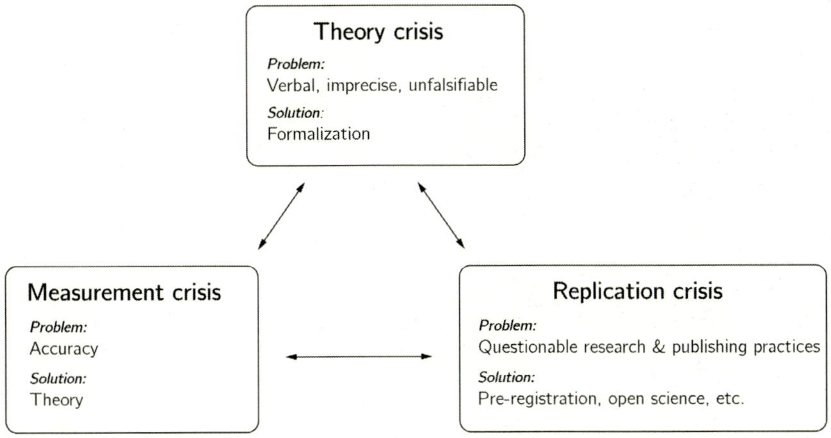

Figure 8.1. The network of crises in psychology. The resolution of one crisis will, I hope, have a positive impact on the resolution of others.

qualitative predictions may be the most feasible outcome. In the previous chapter, I used the example of segregation, which is predicted by many models and their variants, even when individuals are relatively tolerant.

I will not attempt to review these crises in depth, but I would like to suggest, in line with the book, that these crises form a mutualistic network (figure 8.1). Progress in resolving one crisis will have a positive impact on resolving others. If we can have more confidence in the empirical basis of many well-known psychological phenomena, theory development will benefit. Improved theories are necessary to advance measurement, and vice versa. In periods of rapid progress in sciences such as particle physics and biochemistry, we often see an upward spiral of theory development and measurement techniques.

We have been part of at least one such radical transformation, and it is happening now: the AI revolution. Of course, we cannot claim ownership of this revolution, as many disciplines, such as computer science, have played a key role. But the current progress is based on mechanisms (neural learning and operant conditioning) that were first studied by psychologists. Unfortunately, there are no other convincing examples. We have not yet invented the plane or the refrigerator. This is no joke. In science, technology is the proof of the pudding. At some point we really must solve problems like addiction or panic disorder.

Despite all these negatives, psychology remains the most fascinating science of all. As I emphasized in chapter 1, the emergence of global waves of electrical activity from rapid local interactions forms the basis of our conscious thought processes. This phenomenon is truly astonishing, and understanding it presents one of the most intriguing scientific tasks of all time. It involves the ultimate complex system, and we are exploring it with our own minds. Psychology is full of counterintuitive findings and paradoxes. Perhaps the greatest paradox is that psychology, unlike any other science, reveals the limitations and fallibilities of the human intellect while using the very intellect it studies.

The quest to understand the human mind is undeniably one of the most challenging scientific endeavors.

The field of psychology is constantly moving away from grand theories, which were often little more than the collective opinions of some random man, toward more detailed models of subprocesses and systems. Significant progress has been made on a smaller scale. Similarly, the complex-systems approach does not currently provide a grand theory for psychology but, rather, a versatile set of tools for modeling, analysis, and understanding. I cannot conclude with a comprehensive overall theory of the complex human brain–mind–social world system. I simply don't have it, only bits and pieces.

It is my hope that this book will contribute to lasting progress in psychological research. The overview of complex-systems research in other disciplines is perhaps helpful. I also hope that a new generation of researchers will learn many practical, useful modeling skills. And I hope that I have found the unstable maximum between hate and love for psychology.

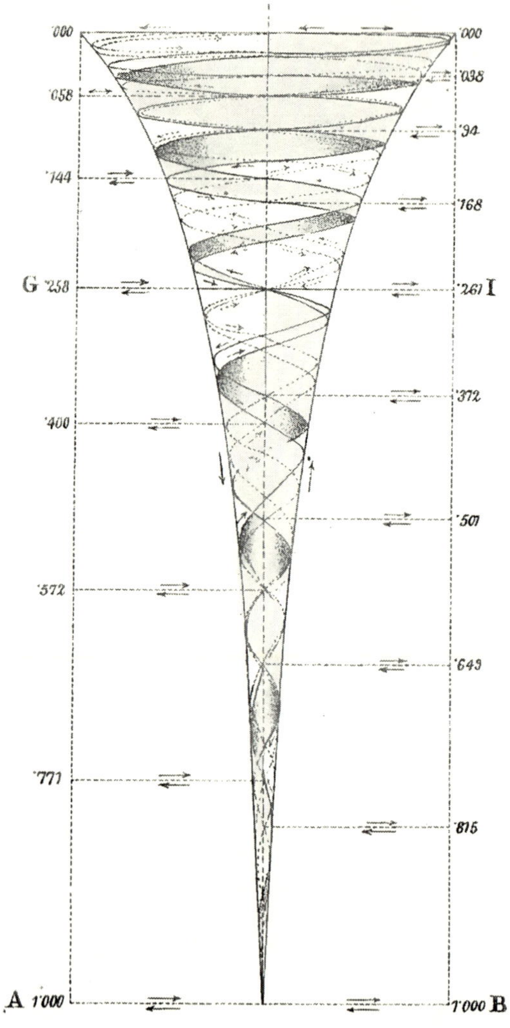

·000
·000
·038
·058
·94
·144
·168
G ·258
·261 I
·372
·400
·501
·572
·649
·771
·815
A 1·000
1·000 B

BIBLIOGRAPHY

Aarts, A. A., J. E. Anderson, C. J. Anderson, P. R. Attridge, A. Attwood, J. Axt, M. Babel, et al. 2015. "Estimating the Reproducibility of Psychological Science." *Science* 349 (6251). https://doi.org/10.1126/science.aac4716.

Abar, S., G. K. Theodoropoulos, P. Lemarinier, and G. M. P. O'Hare. 2017. "Agent Based Modelling and Simulation Tools: A Review of the State-of-Art Software." *Computer Science Review* 24:13–33. https://doi.org/10.1016/j.cosrev.2017.03.001.

Abe, Y., M. Ishida, E. Nozawa, T. Ootsuka, and R. Yahagi. 2017. "Cusp Singularity in Mean Field Ising Model." *European Journal of Physics* 38 (6): 065102. https://doi.org/10.1088/1361-6404/aa82fc.

Abraham, R. 1991. "Cuspoidal Nets." In *Business Cycles: Theories, Evidence and Analysis,* edited by N. Thygesen, K. Velupillai, and S. Zambelli, 56–63. International Economic Association. London, UK: Palgrave Macmillan UK. https://doi.org/10.1007/978-1-349-11570-9_3.

Abraham, R., A. Keith, M. Koebbe, and G. Mayer-Kress. 1991. "Computational Unfolding of Double-Cusp Models of Opinion Formation." *International Journal of Bifurcation and Chaos* 01 (02): 417–430. https://doi.org/10.1142/S0218127491000324.

Abrams, P. A., and L. R. Ginzburg. 2000. "The Nature of Predation: Prey Dependent, Ratio Dependent or Neither?" *Trends in Ecology & Evolution* 15 (8): 337–341. https://doi.org/10.1016/S0169-5347(00)01908-X.

Acebrón, J. A., L. L. Bonilla, C. J. Pérez Vicente, F. Ritort, and R. Spigler. 2005. "The Kuramoto Model: A Simple Paradigm for Synchronization Phenomena." *Reviews of Modern Physics* 77 (1): 137–185. https://doi.org/10.1103/RevModPhys.77.137.

Adams, P., and M. Haire. 1959. "The Effect of Orientation on the Reversal of One Cube Inscribed in Another." *The American Journal of Psychology* 72:296–299. https://doi.org/10.2307/1419384.

Ainsworth, M. D. S., M. C. Blehar, E. Waters, and S. N. Wall. 2015. *Patterns of Attachment: A Psychological Study of the Strange Situation.* New York, NY: Psychology Press. https://doi.org/10.4324/9780203758045.

Alderisio, F., B. G. Bardy, and M. di Bernardo. 2016. "Entrainment and Synchronization in Networks of Rayleigh–van Der Pol Oscillators with Diffusive and Haken–Kelso–Bunz Couplings." *Biological Cybernetics* 110 (2): 151–169. https://doi.org/10.1007/s00422-016-0685-7.

Alexander, R. A., G. R. Herbert, R. P. DeShon, and P. J. Hanges. 1992. "An Examination of Least-Squares Regression Modeling of Catastrophe Theory." *Psychological Bulletin* 111:366–374. https://doi.org/10.1037/0033-2909.111.2.366.

Almaatouq, A., A. Noriega-Campero, A. Alotaibi, P. M. Krafft, M. Moussaid, and A. Pentland. 2020. "Adaptive Social Networks Promote the Wisdom of Crowds." *Proceedings of the National Academy of Sciences* 117 (21): 11379–11386. https://doi.org/10.1073/pnas.1917687117.

Anderson, P. W. 1972. "More Is Different." *Science* 177 (4047): 393–396. https://doi.org/10.1126/science.177.4047.393.

———. 1999a. "Perspective: Complexity Theory and Organization Science." *Organization Science* 10 (3): 216–232. https://doi.org/10.1287/orsc.10.3.216.

———. 1999b. "The Eightfold Way to the Theory of Complexity: A Prologue." In *Complexity: Metaphors, Models, and Reality,* edited by G. Cowan, D. Pines, and D. Meltzer, 7–16. Perseus Books.

Arnaud, M. 2021. *Mixture Modelling from Scratch, in R.* https://towardsdatascience.com/mixture-modelling-from-scratch-in-r-5ab7bfc83eef.

Ashby, W. R. 1956. *An Introduction to Cybernetics.* London, UK: Chapman & Hall Ltd.

Attwell, D., and C. Iadecola. 2002. "The Neural Basis of Functional Brain Imaging Signals." *Trends in Neurosciences* 25 (12): 621–625. https://doi.org/10.1016/S0166-2236(02)02264-6.

Axelrod, R. 1997. "The Dissemination of Culture: A Model with Local Convergence and Global Polarization." *Journal of Conflict Resolution* 41 (2): 203–226. https://doi.org/10.1177/0022002797041002001.

Ayers, S. 1997. "The Application of Chaos Theory to Psychology." *Theory & Psychology* 7:373. https://doi.org/10.1177/0959354397073005.

Bak, P., C. Tang, and K. Wiesenfeld. 1988. "Self-Organized Criticality." *Physical Review A* 38 (1): 364–374. https://doi.org/10.1103/PhysRevA.38.364.

Banks, J., J. Brooks, G. Cairns, G. Davis, and P. Stacey. 1992. "On Devaney's Definition of Chaos." *The American Mathematical Monthly* 99 (4): 332–334. https://doi.org/10.1080/00029890.1992.11995856.

Barabási, A.-L., and M. Pósfai. 2016. *Network Science.* 1st edition. Cambridge, UK: Cambridge University Press.

Barbaro, A. B. T., L. Chayes, and M. R. D'Orsogna. 2013. "Territorial Developments Based on Graffiti: A Statistical Mechanics Approach." *Physica A: Statistical Mechanics and its Applications* 392 (1): 252–270. https://doi.org/10.1016/j.physa.2012.08.001.

Barceló, J., ed. 2010. *Fundamentals of Traffic Simulation.* Vol. 145. International Series in Operations Research & Management Science. New York, NY: Springer. https://doi.org/10.1007/978-1-4419-6142-6.

Barnett, S. M., and S. J. Ceci. 2002. "When and Where Do We Apply What We Learn?: A Taxonomy for Far Transfer." *Psychological Bulletin* 128:612–637. https://doi.org/10.1037/0033-2909.128.4.612.

Barton, S. 1994. "Chaos, Self-Organization, and Psychology." *American Psychologist* 49 (1): 5–14. https://doi.org/10.1037/0003-066X.49.1.5.

Bascompte, J., and P. Jordano. 2013. *Mutualistic Networks.* Princeton, NJ: Princeton University Press.

Battiston, F., E. Amico, A. Barrat, G. Bianconi, G. Ferraz de Arruda, B. Franceschiello, I. Iacopini, et al. 2021. "The Physics of Higher-Order Interactions in Complex Systems." *Nature Physics* 17 (10): 1093–1098. https://doi.org/10.1038/s41567-021-01371-4.

Baumann, F., P. Lorenz-Spreen, I. M. Sokolov, and M. Starnini. 2020. "Modeling Echo Chambers and Polarization Dynamics in Social Networks." *Physical Review Letters* 124 (4): 048301. https://doi.org/10.1103/PhysRevLett.124.048301.

Bechtel, W., and A. Abrahamsen. 2005. "Explanation: A Mechanist Alternative." *Studies in History and Philosophy of Science Part C: Studies in History and Philosophy of Biological and Biomedical Sciences,* Mechanisms in Biology, 36 (2): 421–441. https://doi.org/10.1016/j.shpsc.2005.03.010.

Bédard, C., H. Kröger, and A. Destexhe. 2006. "Does the $1/F$ Frequency Scaling of Brain Signals Reflect Self-Organized Critical States?" *Physical Review Letters* 97 (11): 118102. https://doi.org/10.1103/PhysRevLett.97.118102.

Beek, P. J., and A. Lewbel. 1995. "The Science of Juggling." *Scientific American* 273 (5): 92–97. https://doi.org/10.1038/scientificamerican1195-92.

Beek, P. J., C. E. Peper, and A. Daffertshofer. 2002. "Modeling Rhythmic Interlimb Coordination: Beyond the Haken–Kelso–Bunz Model." *Brain and Cognition* 48 (1): 149–165. https://doi.org/10.1006/brcg.2001.1310.

Bentler, P. M. 1970. "Evidence Regarding Stages in the Development of Conservation." *Perceptual and Motor Skills* 31 (3): 855–859. https://doi.org/10.2466/pms.1970.31.3.855.

Berlekamp, E. R., J. H. Conway, and R. K. Guy. 2004. *Winning Ways for Your Mathematical Plays, Volume 4.* 2nd ed. New York, NY: A. K. Peters/CRC Press. https://doi.org/10.1201/9780429487309.

Bertschinger, N., and T. Natschläger. 2004. "Real-Time Computation at the Edge of Chaos in Recurrent Neural Networks." *Neural Computation* 16 (7): 1413–1436. https://doi.org/10.1162/089976604323057443.

Bizyaeva, A., A. Franci, and N. E. Leonard. 2023. "Nonlinear Opinion Dynamics With Tunable Sensitivity." *IEEE Transactions on Automatic Control* 68 (3): 1415–1430. https://doi.org/10.1109/TAC.2022.3159527.

Blanken, T. F., M. K. Deserno, J. Dalege, D. Borsboom, P. Blanken, G. A. Kerkhof, and A. O. J. Cramer. 2018. "The Role of Stabilizing and Communicating Symptoms given Overlapping Communities in Psychopathology Networks." *Scientific Reports* 8 (1): 5854. https://doi.org/10.1038/s41598-018-24224-2.

Boccaletti, S., G. Bianconi, R. Criado, C. I. del Genio, J. Gómez-Gardeñes, M. Romance, I. Sendiña-Nadal, Z. Wang, and M. Zanin. 2014. "The Structure and Dynamics of Multilayer Networks." *Physics Reports* 544 (1): 1–122. https://doi.org/10.1016/j.physrep.2014.07.001.

Boccaletti, S., V. Latora, Y. Moreno, M. Chavez, and D. Hwang. 2006. "Complex Networks: Structure and Dynamics." *Physics Reports* 424 (4-5): 175–308. https://doi.org/10.1016/j.physrep.2005.10.009.

Boerlijst, M. C., and P. Hogeweg. 1991. "Spiral Wave Structure in Pre-Biotic Evolution: Hypercycles Stable against Parasites." *Physica D: Nonlinear Phenomena* 48 (1): 17–28. https://doi.org/10.1016/0167-2789(91)90049-F.

Bogacz, R., E. Brown, J. Moehlis, P. Holmes, and J. D. Cohen. 2006. "The Physics of Optimal Decision Making: A Formal Analysis of Models of Performance in Two-Alternative Forced-Choice Tasks." *Psychological Review* 113:700–765. https://doi.org/10.1037/0033-295X.113.4.700.

Boostani, R., F. Karimzadeh, and M. Nami. 2017. "A Comparative Review on Sleep Stage Classification Methods in Patients and Healthy Individuals." *Computer Methods and Programs in Biomedicine* 140:77–91. https://doi.org/10.1016/j.cmpb.2016.12.004.

Boot, J., M. van der Ende, R. W. Wiers, M. Lees, and H. L. J. van der Maas. submitted for publication. "Integrating Dual Process Decision-Making and Social Dynamics: A Formal Modeling Framework for Addiction."

Bork, R. V., L. D. Wijsen, and M. Rhemtulla. 2017. "Toward a Causal Interpretation of the Common Factor Model." *Disputatio* 9 (47): 581–601. https://doi.org/10.1515/disp-2017-0019.

Borsboom, D. 2008. "Psychometric Perspectives on Diagnostic Systems." *Journal of Clinical Psychology* 64 (9): 1089–1108. https://doi.org/10.1002/jclp.20503.

———. 2017. "A Network Theory of Mental Disorders." *World Psychiatry* 16 (1): 5–13. https://doi.org/10.1002/wps.20375.

Borsboom, D., A. O. J. Cramer, and A. Kalis. 2019. "Brain Disorders? Not Really: Why Network Structures Block Reductionism in Psychopathology Research." *Behavioral and Brain Sciences* 42:e2. https://doi.org/10.1017/S0140525X17002266.

Borsboom, D., A. O. J. Cramer, V. D. Schmittmann, S. Epskamp, and L. J. Waldorp. 2011. "The Small World of Psychopathology." *PLoS One* 6 (11): e27407. https://doi.org/10.1371/journal.pone.0027407.

Borsboom, D., E. I. Fried, S. Epskamp, L. J. Waldorp, C. D. van Borkulo, H. L. J. van der Maas, and A. O. J. Cramer. 2017. "False Alarm? A Comprehensive Reanalysis of "Evidence That Psychopathology Symptom Networks Have Limited Replicability" by Forbes, Wright, Markon, and Krueger (2017)." *Journal of Abnormal Psychology* 126:989–999. https://doi.org/10.1037/abn0000306.

Borsboom, D., M. Rhemtulla, A. O. J. Cramer, H. L. J. van der Maas, M. Scheffer, and C. V. Dolan. 2016. "Kinds *versus* Continua: A Review of Psychometric Approaches to Uncover the Structure of Psychiatric Constructs." *Psychological Medicine* 46 (8): 1567–1579. https://doi.org/10.1017/S0033291715001944.

Borsboom, D., D. J. Robinaugh, M. Rhemtulla, and A. O. J. Cramer. 2018. "Robustness and Replicability of Psychopathology Networks." *World Psychiatry* 17 (2): 143–144. https://doi.org/10.1002/wps.20515.

Borsboom, D., H. L. J. van der Maas, J. Dalege, R. A. Kievit, and B. D. Haig. 2021. "Theory Construction Methodology: A Practical Framework for Building Theories in Psychology." *Perspectives on Psychological Science* 16 (4): 756–766. https://doi.org/10.1177/17456916 20969647.

Bowles, S., and H. Gintis. 2011. *A Cooperative Species: Human Reciprocity and Its Evolution.* Princeton, NJ: Princeton University Press.

Brandt, M. J. 2022. "Measuring the Belief System of a Person." *Journal of Personality and Social Psychology* 123 (4): 830–853. https://doi.org/10.1037/pspp0000416.

Breakspear, M. 2017. "Dynamic Models of Large-Scale Brain Activity." *Nature Neuroscience* 20 (3): 340–352. https://doi.org/10.1038/nn.4497.

Bringmann, L. F., T. Elmer, S. Epskamp, R. W. Krause, D. Schoch, M. Wichers, J. T. W. Wigman, and E. Snippe. 2019. "What Do Centrality Measures Measure in Psychological Networks?" *Journal of Abnormal Psychology* 128:892–903. https://doi.org/10.1037/abn0000446.

Brinkhuis, M. J. S., Alexander O. Savi, A. D. Hofman, F. Coomans, H. L. J. van der Maas, and G. K. J. Maris. 2018. "Learning As It Happens: A Decade of Analyzing and Shaping a Large-Scale Online Learning System." *Journal of Learning Analytics* 5 (2): 29–46. https://doi.org/10.18608/jla.2018.52.3.

Broer, H., and F. Takens. 2011. *Dynamical Systems and Chaos.* New York, NY: Springer.

Broido, A. D., and A. Clauset. 2019. "Scale-free Networks are Rare." *Nature Communications* 10 (1): 1017. https://doi.org/10.1038/s41467-019-08746-5.

Brush, E. R., D. C. Krakauer, and J. C. Flack. 2018. "Conflicts of Interest Improve Collective Computation of Adaptive Social Structures." *Science Advances* 4 (1): e1603311. https://doi.org/10.1126/sciadv.1603311.

Burger, J., A.-M. Isvoranu, G. Lunansky, J. M. B. Haslbeck, S. Epskamp, R. H. A. Hoekstra, E. I. Fried, D. Borsboom, and T. F. Blanken. 2022. "Reporting Standards for Psychological Network Analyses in Cross-Sectional Data." *Psychological Methods* 28 (4). https://doi.org/10.1037/met0000471.

Busemeyer, J. R., and P. D. Bruza. 2012. *Quantum Models of Cognition and Decision.* Cambridge, UK: Cambridge University Press.

Butler, E. A. 2011. "Temporal Interpersonal Emotion Systems: The "TIES" That Form Relationships." *Personality and Social Psychology Review: An Official Journal of the Society for Personality and Social Psychology, Inc* 15 (4): 367–393. https://doi.org/10.1177/1088868311411164.

Carpenter, G. A., and S. Grossberg. 1987. "A Massively Parallel Architecture for a Self-Organizing Neural Pattern Recognition Machine." *Computer Vision, Graphics, and Image Processing* 37 (1): 54–115. https://doi.org/10.1016/S0734-189X(87)80014-2.

Caspi, A., R. M. Houts, D. W. Belsky, S. J. Goldman-Mellor, H. Harrington, S. Israel, M. H. Meier, et al. 2014. "The p Factor: One General Psychopathology Factor in the Structure of Psychiatric Disorders?" *Clinical Psychological Science : A Journal of the Association for Psychological Science* 2 (2): 119–137. https://doi.org/10.1177/2167702613497473.

Castellano, C., S. Fortunato, and V. Loreto. 2009. "Statistical Physics of Social Dynamics." *Reviews of Modern Physics* 81 (2): 591–646. https://doi.org/10.1103/RevModPhys.81.591.

Castellano, C., M. Marsili, and A. Vespignani. 2000. "Nonequilibrium Phase Transition in a Model for Social Influence." *Physical Review Letters* 85 (16): 3536–3539. https://doi.org/10.1103/PhysRevLett.85.3536.

Castellano, C., M. A. Muñoz, and R. Pastor-Satorras. 2009. "Nonlinear *q*-Voter Model." *Physical Review E* 80 (4): 041129. https://doi.org/10.1103/PhysRevE.80.041129.

Castro, L. N. de, and J. I. Timmis. 2003. "Artificial Immune Systems as a Novel Soft Computing Paradigm." *Soft Computing* 7 (8): 526–544. https://doi.org/10.1007/s00500-002-0237-z.

Cattell, R. B. 1987. *Intelligence: Its Structure, Growth and Action.* Elsevier.

Centola, D. 2022. "The Network Science of Collective Intelligence." *Trends in Cognitive Sciences* 26 (11): 923–941. https://doi.org/10.1016/j.tics.2022.08.009.

Chalmers, D. J. 2006. *Strong and Weak Emergence.* https://philpapers.org/rec/chasaw.

Chambers, C. 2019. *The Seven Deadly Sins of Psychology: A Manifesto for Reforming the Culture of Scientific Practice.* Princeton, NJ: Princeton University Press. https://doi.org/10.1515/9781400884940.

Chambon, M., W. G. Kammeraad, F. van Harreveld, J. Dalege, J. E. Elberse, and H. L. J. van der Maas. 2022. "Understanding Change in COVID-19 Vaccination Intention with Network Analysis of Longitudinal Data from Dutch Adults." *NPJ Vaccines* 7 (1): 1–10. https://doi.org/10.1038/s41541-022-00533-6.

Chang, H. 2008. *Inventing Temperature: Measurement and Scientific Progress.* Oxford Studies in Philosophy of Science. Oxford, UK; New York, NY: Oxford University Press.

Chemero, A. 2013. "Radical Embodied Cognitive Science." *Review of General Psychology* 17 (2): 145–150. https://doi.org/10.1037/a0032923.

Chen, G., and Y. Lou. 2019. *Naming Game: Models, Simulations and Analysis.* Vol. 34. Emergence, Complexity and Computation. Cham: Springer International Publishing. https://doi.org/10.1007/978-3-030-05243-0.

Chialvo, D. R. 2010. "Emergent Complex Neural Dynamics." *Nature Physics* 6 (10): 744–750. https://doi.org/10.1038/nphys1803.

Christensen, A. P., and Y. N. Kenett. 2021. "Semantic Network Analysis (SemNA): A Tutorial on Preprocessing, Estimating, and Analyzing Semantic Networks." *Psychological Methods* 28 (4): 860–879. https://doi.org/10.1037/met0000463.

Clark, A. 2013. "Whatever next? Predictive Brains, Situated Agents, and the Future of Cognitive Science." *Behavioral and Brain Sciences* 36 (3): 181–204. https://doi.org/10.1017/S0140525X12000477.

Clark, D. M. 1986. "A Cognitive Approach to Panic." *Behaviour Research and Therapy* 24 (4): 461–470. https://doi.org/10.1016/0005-7967(86)90011-2.

Clauset, A., C. Moore, and M. E. J. Newman. 2008. "Hierarchical Structure and the Prediction of Missing Links in Networks." *Nature* 453 (7191): 98–101. https://doi.org/10.1038/nature06830.

Clayton, P., and P. Davies, eds. 2008. *The Re-Emergence of Emergence.* Oxford University Press. https://doi.org/10.1093/acprof:oso/9780199544318.001.0001.

Cleeremans, A., D. Achoui, A. Beauny, L. Keuninckx, J.-R. Martin, S. Muñoz-Moldes, L. Vuillaume, and A. de Heering. 2020. "Learning to Be Conscious." *Trends in Cognitive Sciences* 24 (2): 112–123. https://doi.org/10.1016/j.tics.2019.11.011.

Cobb, L. 1978. "Stochastic Catastrophe Models and Multimodal Distributions." *Behavioral Science* 23 (4): 360–374. https://doi.org/10.1002/bs.3830230407.

Cobb, L., and S. Zacks. 1985. "Applications of Catastrophe Theory for Statistical Modeling in the Biosciences." *Journal of the American Statistical Association* 80 (392): 793–802. https://doi.org/10.1080/01621459.1985.10478184.

Cocchi, L., L. L. Gollo, A. Zalesky, and M. Breakspear. 2017. "Criticality in the Brain: A Synthesis of Neurobiology, Models and Cognition." *Progress in Neurobiology* 158:132–152. https://doi.org/10.1016/j.pneurobio.2017.07.002.

Cohen, A., D. Pargman, and G. Tenenbaum. 2003. "Critical Elaboration and Empirical Investigation of the Cusp Catastrophe Model: A Lesson for Practitioners." *Journal of Applied Sport Psychology* 15 (2): 144–159. https://doi.org/10.1080/10413200305393.

Costantini, G., S. Epskamp, D. Borsboom, M. Perugini, R. Mõttus, L. J. Waldorp, and A. O. J. Cramer. 2015. "State of the aRt Personality Research: A Tutorial on Network Analysis of Personality Data in R." *Journal of Research in Personality,* R Special Issue, 54:13–29. https://doi.org/10.1016/j.jrp.2014.07.003.

Cramer, A. O. J., C. D. van Borkulo, E. J. Giltay, H. L. J. van der Maas, K. S. Kendler, M. Scheffer, and D. Borsboom. 2016. "Major Depression as a Complex Dynamic System." *PLoS One* 11 (12): e0167490. https://doi.org/10.1371/journal.pone.0167490.

Cramer, A. O. J., S. Van Der Sluis, A. Noordhof, M. Wichers, N. Geschwind, S. H. Aggen, K. S. Kendler, and D. Borsboom. 2012. "Dimensions of Normal Personality as Networks in Search of Equilibrium: You Can't like Parties if you Don't like People." *European Journal of Personality* 26 (4): 414–431. https://doi.org/10.1002/per.1866.

Cramer, A. O. J., L. J. Waldorp, H. L. J. van der Maas, and D. Borsboom. 2010. "Comorbidity: A Network Perspective." *The Behavioral and Brain Sciences* 33 (2-3): 137–150, discussion 150–193. https://doi.org/10.1017/S0140525X09991567.

Crielaard, L., J. F. Uleman, B. D. L. Châtel, S. Epskamp, P. M. A. Sloot, and R. Quax. 2022. "Refining the Causal Loop Diagram: A Tutorial for Maximizing the Contribution of Domain Expertise in Computational System Dynamics Modeling." *Psychological Methods* 29 (1): 169–201. https://doi.org/10.1037/met0000484.

Cronbach, L. J. 1957. "The Two Disciplines of Scientific Psychology." *American Psychologist* 12:671–684. https://doi.org/10.1037/h0043943.

da Mata, A. S. 2020. "Complex Networks: A Mini-review." *Brazilian Journal of Physics* 50 (5): 658–672. https://doi.org/10.1007/s13538-020-00772-9.

Dablander, F., and M. Hinne. 2019. "Node Centrality Measures Are a Poor Substitute for Causal Inference." *Scientific Reports* 9 (1): 6846. https://doi.org/10.1038/s41598-019-43033-9.

Dablander, F., A. Pichler, A. Cika, and A. Bacilieri. 2023. "Anticipating Critical Transitions in Psychological Systems Using Early Warning Signals: Theoretical and Practical Considerations." *Psychological Methods* 28 (4): 765–790. https://doi.org/10.1037/met0000450.

Dablander, F., and R. van Bork. 2021. "Causal Inference." In *Network Psychometrics with R*, edited by A. M. Isvoranu, S. Eskamp, L. J. Waldrop, and D. Borsboom, 213–232. Abingdon, VA; Oxford, UK: Routledge.

Dakos, V., S. R. Carpenter, W. A. Brock, A. M. Ellison, V. Guttal, A. R. Ives, S. Kéfi, et al. 2012. "Methods for Detecting Early Warnings of Critical Transitions in Time Series Illustrated Using Simulated Ecological Data." *PloS One* 7 (7): e41010. https://doi.org/10.1371/journal.pone.0041010.

Dalege, J. 2020. "A Formal Approach to Attitude." PhD diss., Universiteit van Amsterdam. https://api.semanticscholar.org/CorpusID:213884605.

Dalege, J., D. Borsboom, F. van Harreveld, H. van den Berg, M. Conner, and H. L. J. van der Maas. 2016. "Toward a Formalized Account of Attitudes: The Causal Attitude Network (CAN) Model." *Psychological Review* 123 (1): 2–22. https://doi.org/10.1037/a0039802.

Dalege, J., D. Borsboom, F. van Harreveld, and H. L. J. van der Maas. 2017. "Network Analysis on Attitudes: A Brief Tutorial." *Social Psychological and Personality Science* 8 (5): 528–537. https://doi.org/10.1177/1948550617709827.

———. 2018. "The Attitudinal Entropy (AE) Framework as a General Theory of Individual Attitudes." *Psychological Inquiry* 29 (4): 175–193. https://doi.org/10.1080/1047840X.2018.1537246.

Dalege, J., and H. L. J. van der Maas. 2020. "Accurate by Being Noisy: A Formal Network Model of Implicit Measures of Attitudes." *Social Cognition* 38:S26–S41. https://doi.org/10.1521/soco.2020.38.supp.s26.

Dandekar, P., A. Goel, and D. T. Lee. 2013. "Biased Assimilation, Homophily, and the Dynamics of Polarization." *Proceedings of the National Academy of Sciences* 110 (15): 5791–5796. https://doi.org/10.1073/pnas.1217220110.

Dattagupta, S. 1981. "The Tricritical Point—A Qualitative Review." *Bulletin of Material Science* 3 (2): 133–139. https://doi.org/10.1007/BF02908488.

Daugherty, D., T. Roque-Urrea, J. Urrea-Roque, J. Troyer, S. Wirkus, and M. A. Porter. 2009. "Mathematical Models of Bipolar Disorder." *Communications in Nonlinear Science and Numerical Simulation* 14 (7): 2897–2908. https://doi.org/10.1016/j.cnsns.2008.10.027.

Dawkins, R. 1986. *The Blind Watchmaker.* New York, NY: New York : Norton.

de Boer, R. J. 2018. "Simple Phase Plane Analysis and Parameter Estimation in R," no. 20, https://tbb.bio.uu.nl/khwjtuss/ProtophloemModel/AuxinAux1SingleCellModel/tutorial.pdf.

De Groot, S. R., and P. Mazur. 2013. *Non-Equilibrium Thermodynamics.* Courier Corporation.

de Mooij, S. M. M., T. F. Blanken, R. P. P. P. Grasman, J. R. Ramautar, E. J. W. Van Someren, and H. L. J. van der Maas. 2020. "Dynamics of Sleep: Exploring Critical Transitions and Early Warning Signals." *Computer Methods and Programs in Biomedicine* 193:105448. https://doi.org/10.1016/j.cmpb.2020.105448.

de Ron, J., M. Deserno, D. J. Robinaugh, D. Borsboom, and H. L. J. van der Maas. 2023. "Towards a General Modeling Framework of Resource Competition in Cognitive Development." *Child Development* 94 (6): 1432–1453. https://doi.org/10.1111/cdev.13973.

Deffuant, G., D. Neau, F. Amblard, and G. Weisbuch. 2000. "Mixing Beliefs among Interacting Agents." *Advances in Complex Systems* 03 (01n04): 87–98. https://doi.org/10.1142/S0219525900000078.

Dekker, M. M., A. S. von der Heydt, and H. A. Dijkstra. 2018. "Cascading Transitions in the Climate System." *Earth System Dynamics* 9 (4): 1243–1260. https://doi.org/10.5194/esd-9-1243-2018.

DellaPosta, D. 2020. "Pluralistic Collapse: The "Oil Spill" Model of Mass Opinion Polarization." *American Sociological Review* 85 (3): 507–536. https://doi.org/10.1177/0003122420922989.

Den Hartigh, R. J. R., Marijn W. G. Van Dijk, H. W. Steenbeek, and P. L. C. Van Geert. 2016. "A Dynamic Network Model to Explain the Development of Excellent Human Performance." *Frontiers in Psychology* 7. https://doi.org/10.3389/fpsyg.2016.00532.

Deserno, M. K., D. Borsboom, S. Begeer, R. van Bork, M. Hinne, and H. M. Geurts. 2020. "Highways to Happiness for Autistic Adults? Perceived Causal Relations among Clinicians." *PLoS One* 15 (12): e0243298. https://doi.org/10.1371/journal.pone.0243298.

Dodson, M. M., and A. Hallam. 1977. "Allopatric Speciation and the Fold Catastrophe." *The American Naturalist* 111 (979): 415–433. https://doi.org/10.1086/283176.

Doerig, A., A. Schurger, and M. H. Herzog. 2021. "Hard Criteria for Empirical Theories of Consciousness." *Cognitive Neuroscience* 12 (2): 41–62. https://doi.org/10.1080/17588928.2020.1772214.

Dolan, C. V., and H. L. J. van der Maas. 1998. "Fitting Multivariage Normal Finite Mixtures Subject to Structural Equation Modeling." *Psychometrika* 63 (3): 227–253. https://doi.org/10.1007/BF02294853.

Dong, Y., M. Zhan, G. Kou, Z. Ding, and H. Liang. 2018. "A Survey on the Fusion Process in Opinion Dynamics." *Information Fusion* 43:57–65. https://doi.org/10.1016/j.inffus.2017.11.009.

Dorogovtsev, S. N., and J. F. F. Mendes. 2002. "Evolution of Networks." *Advances in Physics* 51 (4): 1079–1187. https://doi.org/10.1080/00018730110112519.

Douglas, K. M., R. M. Sutton, and Aleksandra Cichocka. 2017. "The Psychology of Conspiracy Theories." *Current Directions in Psychological Science* 26:538–542. https://doi.org/10.1177/0963721417718261.

Dresp-Langley, B. 2020. "Seven Properties of Self-Organization in the Human Brain." *Big Data and Cognitive Computing* 2 (4): 10. https://doi.org/10.3390/bdcc4020010.

Duckworth, A., and J. J. Gross. 2014. "Self-Control and Grit: Related but Separable Determinants of Success." *Current Directions in Psychological Science* 23 (5): 319–325. https://doi.org/10.1177/0963721414541462.

Duckworth, A. L., and M. L. Kern. 2011. "A Meta-Analysis of the Convergent Validity of Self-Control Measures." *Journal of Research in Personality* 45 (3): 259–268. https://doi.org/10.1016/j.jrp.2011.02.004.

Durlauf, S. N., and H. P. Young. 2001. *Social Dynamics.* Cambridge, MA: MIT Press.

Dutilh, G., E.-J. Wagenmakers, I. Visser, and H. L. J. van der Mass. 2011. "A Phase Transition Model for the Speed-Accuracy Trade-Off in Response Time Experiments." *Cognitive Science* 35 (2): 211–250. https://doi.org/10.1111/j.1551-6709.2010.01147.x.

Ebaugh, H. R. F. 1988. *Becoming an Ex: The Process of Role Exit.* Chicago, IL: University of Chicago Press.

Edelman, G. M. 1987. *Neural Darwinism: The Theory of Neuronal Group Selection.* xxii, 371. New York, NY: Basic Books.

Edelmann, A., T. Wolff, D. Montagne, and C. A. Bail. 2020. "Computational Social Science and Sociology." *Annual Review of Sociology* 46 (1): 61–81. https://doi.org/10.1146/annurev-soc-121919-054621.

Edwards, J. R., and R. P. Bagozzi. 2000. "On the Nature and Direction of Relationships between Constructs and Measures." *Psychological Methods* 5:155–174. https://doi.org/10.1037/1082-989X.5.2.155.

Eigen, M., and P. Schuster. 1979. *The Hypercycle.* Berlin, Heidelberg: Springer. https://doi.org/10.1007/978-3-642-67247-7.

Eldredge, N., and S. J. Gould. 1972. "Punctuated Equilibria: An Alternative to Phyletic Gradualism." In *Models in Paleobiology,* edited by T. J. M. Schopf, 82–115. Freeman Cooper.

Elo, A. E. 1978. *The Rating of Chessplayers, Past and Present.* New York, NY: Arco Pub.

Elovainio, M., J. Lipsanen, L. Pulkki-Råback, J. Suvisaari, and C. Hakulinen. 2021. "Is Symptom Connectivity Really the Most Important Issue in Depression? Depression as a Dynamic System of Interconnected Symptoms Revisited." *Journal of Psychiatric Research* 142:250–257. https://doi.org/10.1016/j.jpsychires.2021.08.004.

Epskamp, S., D. Borsboom, and E. I. Fried. 2018. "Estimating Psychological Networks and Their Accuracy: A Tutorial Paper." *Behavior Research Methods* 50 (1): 195–212. https://doi.org/10.3758/s13428-017-0862-1.

Epskamp, S., A. O. J. Cramer, L. J. Waldorp, V. D. Schmittmann, and D. Borsboom. 2012. "Qgraph: Network Visualizations of Relationships in Psychometric Data." *Journal of Statistical Software* 48:1–18. https://doi.org/10.18637/jss.v048.i04.

Epskamp, S., M. Rhemtulla, and D. Borsboom. 2017. "Generalized Network Psychometrics: Combining Network and Latent Variable Models." *Psychometrika* 82 (4): 904–927. https://doi.org/10.1007/s11336-017-9557-x.

Epskamp, S., H. L. J. van der Maas, R. E. Peterson, H. M. van Loo, S. H. Aggen, and K. S. Kendler. 2022. "Intermediate Stable States in Substance Use." *Addictive Behaviors* 129:107252. https://doi.org/10.1016/j.addbeh.2022.107252.

Epskamp, S., L. J. Waldorp, R. Mõttus, and D. Borsboom. 2018. "The Gaussian Graphical Model in Cross-Sectional and Time-Series Data." *Multivariate Behavioral Research* 53 (4): 453–480. https://doi.org/10.1080/00273171.2018.1454823.

Epstein, J. M. 2007. *Generative Social Science: Studies in Agent-Based Computational Modeling.* Princeton, NJ: Princeton University Press.

———. 2013. *Agent_Zero: Toward Neurocognitive Foundations for Generative Social Science.* Princeton, NJ: Princeton University Press.

Epstein, J. M., and R. Axtell. 1996. *Growing Artificial Societies: Social Science from the Bottom Up.* xv, 208. Cambridge, MA: The MIT Press.

Eronen, M. I., and L. F. Bringmann. 2021. "The Theory Crisis in Psychology: How to Move Forward." *Perspectives on Psychological Science* 16 (4): 779–788. https://doi.org/10.1177/1745691620970586.

Favela, L. H., and M. M. J. W. van Rooij. 2019. "Reasoning across Continuous Landscapes: A Nonlinear Dynamical Systems Theory Approach to Reasoning." *Cognitive Systems Research* 54:189–198. https://doi.org/10.1016/j.cogsys.2018.12.013.

Ferri, I., A. Díaz-Guilera, and M. Palassini. 2022. *Equilibrium and Dynamics of a Three-State Opinion Model.* https://doi.org/10.48550/arXiv.2210.03054.

Festinger, L. 1962. *A Theory of Cognitive Dissonance.* xi, 291. Palo Alto, CA: Stanford Univer. Press.

Finnemann, A., D. Borsboom, S. Epskamp, and H. L. J. van der Maas. 2021. "The Theoretical and Statistical Ising Model: A Practical Guide in R." *Psych* 3 (4): 593–617. https://doi.org/10.3390/psych3040039.

Flache, A., M. Mäs, T. Feliciani, E. Chattoe-Brown, G. Deffuant, S. Huet, and J. Lorenz. 2017. "Models of Social Influence: Towards the Next Frontiers." *Journal of Artificial Societies and Social Simulation* 20 (4): 2. https://doi.org/10.18564/jasss.3521.

Flack, J. C. 2017. "Coarse-Graining as a Downward Causation Mechanism." *Philosophical Transactions of the Royal Society A: Mathematical, Physical and Engineering Sciences* 375 (2109): 20160338. https://doi.org/10.1098/rsta.2016.0338.

Fodor, J. A. 1974. "Special Sciences (or: The Disunity of Science as a Working Hypothesis)." *Synthese* 28 (2): 97–115. https://doi.org/10.1007/BF00485230.

Forbes, M. K., A. G. C. Wright, K. E. Markon, and R. F. Krueger. 2017. "Evidence That Psychopathology Symptom Networks Have Limited Replicability." *Journal of Abnormal Psychology* 126:969–988. https://doi.org/10.1037/abn0000276.

Forger, D. B., M. E. Jewett, and R. E. Kronauer. 1999. "A Simpler Model of the Human Circadian Pacemaker." *Journal of biological rhythms* 14 (6): 532–537. https://doi.org/10.1177/074873099129000867.

Forrester, J. W. 1993. "System Dynamics and the Lessons of 35 Years." In *A Systems-Based Approach to Policymaking,* edited by K. B. De Greene, 199–240. Boston, MA: Springer US. https://doi.org/10.1007/978-1-4615-3226-2_7.

Fortmann-Roe, S. 2014. "Insight Maker: A General-Purpose Tool for Web-Based Modeling & Simulation." *Simulation Modelling Practice and Theory* 47:28–45. https://doi.org/10.1016/j.simpat.2014.03.013.

Fortunato, S. 2004. "Universality of the Threshold for Complete Consensus for the Opinion Dynamics of Deffuant." *International Journal of Modern Physics C* 15:1301–1307. https://doi.org/10.1142/S0129183104006728.

Fortunato-Tavares, T., C. R. F. de Andrade, D. M. Befi-Lopes, A. Hestvik, B. Epstein, L. Tornyova, and R. G. Schwartz. 2012. "Syntactic Structural Assignment in Brazilian Portuguese-Speaking Children with Specific Language Impairment." *Journal of Speech, Language, and Hearing Research* 55 (4): 1097–1111. http://search.ebscohost.com/login.aspx?direct=true&AuthType=ip,sso&db=eric&AN=EJ984113&site=ehost-live&custid=s1098328.

Fowler, J. H. 2005. "Second-Order Free-Riding Problem Solved?" *Nature* 437 (7058): E8–E8. https://doi.org/10.1038/nature04201.

Franz, D. J. 2022. ""Are Psychological Attributes Quantitative?" Is Not an Empirical Question: Conceptual Confusions in the Measurement Debate." *Theory & Psychology* 32 (1): 131–150. https://doi.org/10.1177/09593543211045340.

Fraser, S. 1995. *The Bell Curve Wars: Race, Intelligence, and the Future of America.* Basic Books.

Freitas, U., E. Roulin, J.-F. Muir, and C. Letellier. 2009. "Identifying Chaos from Heart Rate: The Right Task?" *Chaos: An Interdisciplinary Journal of Nonlinear Science* 19 (2): 028505. https://doi.org/10.1063/1.3139116.

Fried, E. I., and A. O. J. Cramer. 2017. "Moving Forward: Challenges and Directions for Psychopathological Network Theory and Methodology." *Perspectives on Psychological Science* 12 (6): 999–1020. https://doi.org/10.1177/1745691617705892.

Fried, E. I., S. Epskamp, R. M. Nesse, F. Tuerlinckx, and D. Borsboom. 2016. "What Are 'good' Depression Symptoms? Comparing the Centrality of DSM and Non-DSM Symptoms of Depression in a Network Analysis." *Journal of Affective Disorders* 189:314–320. https://doi.org/10.1016/j.jad.2015.09.005.

Fried, E. I., C. D. van Borkulo, A. O. J. Cramer, L. Boschloo, R. A. Schoevers, and D. Borsboom. 2017. "Mental Disorders as Networks of Problems: A Review of Recent Insights." *Social Psychiatry and Psychiatric Epidemiology* 52 (1): 1–10. https://doi.org/10.1007/s00127-016-1319-z.

Friedkin, N. E., and E. C. Johnsen. 1990. "Social Influence and Opinions." *The Journal of Mathematical Sociology* 15 (3-4): 193–206. https://doi.org/10.1080/0022250X.1990.9990069.

Friedman, J., T. Hastie, and R. Tibshirani. 2008. "Sparse Inverse Covariance Estimation with the Graphical Lasso." *Biostatistics* 9 (3): 432–441. https://doi.org/10.1093/biostatistics/kxm045.

Friston, K. 2009. "The Free-Energy Principle: A Rough Guide to the Brain?" *Trends in Cognitive Sciences* 13 (7): 293–301. https://doi.org/10.1016/j.tics.2009.04.005.

Fuchs, A., and S. Kelso. 2018. "Coordination Dynamics and Synergetics: From Finger Movements to Brain Patterns and Ballet Dancing." In *Complexity and Synergetics,* 301–316. https://doi.org/10.1007/978-3-319-64334-2_23.

Fürstenau, N. 2014. "Simulating Bistable Perception with Interrupted Ambiguous Stimulus Using Self-Oscillator Dynamics with Percept Choice Bifurcation." *Cognitive Processing* 15 (4): 467–490. https://doi.org/10.1007/s10339-014-0630-4.

Fytas, N. G., V. Martin-Mayor, M. Picco, and N. Sourlas. 2018. "Review of Recent Developments in the Random-Field Ising Model." *Journal of Statistical Physics* 172 (2): 665–672. https://doi.org/10.1007/s10955-018-1955-7.

Galam, S. 2008. "Sociophysics: A Review of Galam Models." *International Journal of Modern Physics C* 19 (03): 409–440. https://doi.org/10.1142/S0129183108012297.

Galam, S., Y. Gefen, and Y. Shapir. 1982. "Sociophysics: A New Approach of Sociological Collective Behaviour. I. Mean-behaviour Description of a Strike." *The Journal of Mathematical Sociology* 9 (1): 1–13. https://doi.org/10.1080/0022250X.1982.9989929.

Gale, D., and L. S. Shapley. 1962. "College Admissions and the Stability of Marriage." *The American Mathematical Monthly* 69 (1): 9–15. https://doi.org/10.2307/2312726.

Galesic, M., D. Barkoczi, A. M. Berdahl, D. Biro, G. Carbone, I. Giannoccaro, R. L. Goldstone, et al. 2023. "Beyond Collective Intelligence: Collective Adaptation." *Journal of The Royal Society Interface* 20 (200): 20220736. https://doi.org/10.1098/rsif.2022.0736.

Galesic, M., and D. L. Stein. 2019. "Statistical Physics Models of Belief Dynamics: Theory and Empirical Tests." *Physica A: Statistical Mechanics and its Applications* 519:275–294. https://doi.org/10.1016/j.physa.2018.12.011.

Galton, F. 1907. "Vox Populi." *Nature* 75 (1949): 450–451. https://doi.org/10.1038/075450a0.

Garcez, A. d'A., and L. C. Lamb. 2023. "Neurosymbolic AI: the 3rd wave." *Artificial Intelligence Review* 56 (11): 12387–12406. https://doi.org/10.1007/s10462-023-10448-w.

Garcez, Artur S. D'Avila, Luís C. Lamb, and Dov M. Gabbay. 2009. *Neural-Symbolic Cognitive Reasoning.* Berlin: Springer.

Gärtner, K. 1990. "A Third Component Causing Random Variability beside Environment and Genotype. A Reason for the Limited Success of a 30 Year Long Effort to Standardize Laboratory Animals?" *Laboratory Animals* 24 (1): 71–77. https://doi.org/10.1258/002367790780890347.

Gauvin, L., J. Vannimenus, and J.-P. Nadal. 2009. "Phase Diagram of a Schelling Segregation Model." *The European Physical Journal B* 70 (2): 293–304. https://doi.org/10.1140/epjb/e2009-00234-0.

Gell-Mann, M. 2002. "What Is Complexity?" In *Complexity and Industrial Clusters,* edited by A. Q. Curzio and M. Fortis, 13–24. Contributions to Economics. Heidelberg: Physica-Verlag HD. https://doi.org/10.1007/978-3-642-50007-7_2.

Ghatak, A. 2019. *Deep Learning with R.* Singapore: Springer. https://doi.org/10.1007/978-981-13-5850-0.

Gibson, J. J. 2014. *The Ecological Approach to Visual Perception: Classic Edition.* New York, NY: Psychology Press. https://doi.org/10.4324/9781315740218.

Gilmore, R. 1993. *Catastrophe Theory for Scientists and Engineers.* New York, NY: Dover.

Goldstone, R. L., and M. A. Janssen. 2005. "Computational Models of Collective Behavior." *Trends in Cognitive Sciences* 9 (9): 424–430. https://doi.org/10.1016/j.tics.2005.07.009.

Gottman, J. M., and R. W. Levenson. 1992. "Marital Processes Predictive of Later Dissolution: Behavior, Physiology, and Health." *Journal of Personality and Social Psychology* 63:221–233. https://doi.org/10.1037/0022-3514.63.2.221.

Gottman, J. M., J. D. Murray, C. C. Swanson, R. Tyson, and K. R. Swanson. 2002. *The Mathematics of Marriage: Dynamic Nonlinear Models.* xvii, 403. Cambridge, MA: MIT Press.

Granovetter, M. 1973. "The Strength of Weak Ties." *American Journal of Sociology* 78 (6): 1360–1380. https://doi.org/10.1086/225469.

———. 1978. "Threshold Models of Collective Behavior." *The American Journal of Sociology* 83 (6): 1420–1443. https://doi.org/10.1086/226707.

Grasman, R., H. L. J. van der Maas, and E.-J. Wagenmakers. 2009. "Fitting the Cusp Catastrophe in R: A cusp Package Primer." *Journal of Statistical Software* 32 (8). https://ideas.repec.org/a/jss/jstsof/v032i08.html.

Grauwin, S., G. Beslon, É. Fleury, S. Franceschelli, C. Robardet, J.-B. Rouquier, and P. Jensen. 2012. "Complex Systems Science: Dreams of Universality, Interdisciplinarity Reality." *Journal of the American Society for Information Science and Technology* 63 (7): 1327–1338. https://doi.org/10.1002/asi.22644.

Grossberg, S., and B. Pinna. 2012. "Neural Dynamics of Gestalt Principles of Perceptual Organization: From Grouping to Shape and Meaning." *Gestalt Theory* 34:399–482.

Guastello, S., A. Correro, and D. Marra. 2019. "Cusp Catastrophe Models for Cognitive Workload and Fatigue in Teams." *Applied Ergonomics* 79:152–168. https://doi.org/10.1016/j.apergo.2018.08.019.

Guastello, S. J. 1982. "Moderator Regression and the Cusp Catastrophe: Application of Two-Stage Personnel Selection, Training, Therapy, and Policy Evaluation." *Behavioral Science* 27 (3): 259–272. https://doi.org/10.1002/bs.3830270305.

———. 1984. "Cusp and Butterfly Catastrophe Modeling of Two Opponent Process Models: Drug Addiction and Work Performance." *Behavioral Science* 29 (4): 258–262. https://doi.org/10.1002/bs.3830290405.

Guastello, S. J., M. Koopmans, and D. Pincus. 2008. *Chaos and Complexity in Psychology: The Theory of Nonlinear Dynamical Systems.* Cambridge University Press.

Guckenheimer, J., and P. Holmes. 1983. "Global Bifurcations." In *Nonlinear Oscillations, Dynamical Systems, and Bifurcations of Vector Fields,* edited by J. Guckenheimer and P. Holmes, 289–352. Applied Mathematical Sciences. New York, NY: Springer. https://doi.org/10.1007/978-1-4612-1140-2_6.

Güémez, J., C. Fiolhais, and M. Fiolhais. 2002. "The Cartesian Diver and the Fold Catastrophe." *American Journal of Physics* 70 (7): 710–714. https://doi.org/10.1119/1.1477433.

Guimerà, R., B. Uzzi, J. Spiro, and L. A. N. Amaral. 2005. "Team Assembly Mechanisms Determine Collaboration Network Structure and Team Performance." *Science* 308 (5722): 697–702. https://doi.org/10.1126/science.1106340.

Haig, B. D. 2005. "An Abductive Theory of Scientific Method." *Psychological Methods* 10 (4): 371–388. https://doi.org/10.1037/1082-989X.10.4.371.

———. 2014. *Investigating the Psychological World: Scientific Method in the Behavioral Sciences.* MIT Press.

Haken, H. 1977. "Synergetics." *Physics Bulletin* 28 (9): 412. https://doi.org/10.1088/0031-9112/28/9/027.

———. 1992. "Synergetics in Psychology." In *Self-Organization and Clinical Psychology: Empirical Approaches to Synergetics in Psychology,* edited by W. Tschacher, G. Schiepek, and E. J. Brunner, 32–54. Springer Series in Synergetics. Berlin, Heidelberg: Springer. https://doi.org/10.1007/978-3-642-77534-5_2.

Haken, H., J. A. S. Kelso, and H. Bunz. 1985. "A Theoretical Model of Phase Transitions in Human Hand Movements." *Biological Cybernetics* 51 (5): 347–356. https://doi.org/10.1007/BF00336922.

Hamaker, E. L., E. Ceulemans, R. P. P. P. Grasman, and F. Tuerlinckx. 2015. "Modeling Affect Dynamics: State of the Art and Future Challenges." *Emotion Review* 7 (4): 316–322. https://doi.org/10.1177/1754073915590619.

Hamaker, E. L., Z. Zhang, and H. L. J. van der Maas. 2009. "Using Threshold Autoregressive Models to Study Dyadic Interactions." *Psychometrika* 74 (4): 727–745. https://doi.org/10.1007/s11336-009-9113-4.

Hameroff, S., and R. Penrose. 1996. "Orchestrated Reduction of Quantum Coherence in Brain Microtubules: A Model for Consciousness." *Mathematics and Computers in Simulation* 40 (3): 453–480. https://doi.org/10.1016/0378-4754(96)80476-9.

Hardin, G. 1968. "The Tragedy of the Commons." *Science* 162 (3859): 1243–1248. https://www.jstor.org/stable/1724745.

Hardy, L. 1996. "Testing the Predictions of the Cusp Catastrophe Model of Anxiety and Performance." *The Sport Psychologist* 10 (2): 140–156. https://doi.org/10.1123/tsp.10.2.140.

Hardy, L., and G. Parfitt. 1991. "A catastrophe model of anxiety and performance." *British Journal of Psychology* 82 (2): 163–178. https://doi.org/10.1111/j.2044-8295.1991.tb02391.x.

Hardy, L., T. Woodman, and S. Carrington. 2004. "Is Self-Confidence a Bias Factor in Higher Order Catastrophe Models? An Exploratory Analysis." *Journal of Sport and Exercise Psychology* 26 (3): 359–368. https://doi.org/10.1123/jsep.26.3.359.

Harezlak, K., and P. Kasprowski. 2018. "Searching for Chaos Evidence in Eye Movement Signals." *Entropy* 20 (1): 32. https://doi.org/10.3390/e20010032.

Haslbeck, J. M. B., and O. Ryan. 2022. "Recovering Within-Person Dynamics from Psychological Time Series." *Multivariate Behavioral Research* 57 (5): 735–766. https://doi.org/10.1080/00273171.2021.1896353.

Haslbeck, J. M. B., O. Ryan, D. J. Robinaugh, L. J. Waldorp, and D. Borsboom. 2022. "Modeling Psychopathology: From Data Models to Formal Theories." *Psychological Methods* 27:930–957. https://doi.org/10.1037/met0000303.

Hassell, M. P., J. H. Lawton, and R. M. May. 1976. "Patterns of Dynamical Behaviour in Single-Species Populations." *Journal of Animal Ecology* 45 (2): 471–486. https://doi.org/10.2307/3886.

Hayes, A. M., and L. A. Andrews. 2020. "A Complex Systems Approach to the Study of Change in Psychotherapy." *BMC Medicine* 18 (1): 197. https://doi.org/10.1186/s12916-020-01662-2.

Heath, R. A. 2000. *Nonlinear Dynamics: Techniques and Applications in Psychology.* 1st edition. Mahwah, NJ: Psychology Press.

Hebb, D. O. 1949. *The Organization of Behavior; a Neuropsychological Theory.* xix, 335. Oxford, UK: Wiley.

Heckman, J. J. 1995. "Lessons from the Bell Curve." *Journal of Political Economy* 103 (5): 1091–1120. https://doi.org/10.1086/262014.

Hegselmann, R. 2017. "Thomas C. Schelling and James M. Sakoda: The Intellectual, Technical, and Social History of a Model." *Journal of Artificial Societies and Social Simulation* 20 (3): 15. https://doi.org/10.18564/jasss.3511.

Hegselmann, R., and U. Krause. 2002. "Opinion Dynamics and Bounded Confidence: Models, Analysis and Simulation." *Journal of Artificial Societies and Social Simulation* 5 (3). https://jasss.soc.surrey.ac.uk/5/3/2.html.

Heider, F. 1946. "Attitudes and Cognitive Organization." *The Journal of Psychology* 21 (1): 107–112. https://doi.org/10.1080/00223980.1946.9917275.

Helbing, D., and P. Molnár. 1995. "Social Force Model for Pedestrian Dynamics." *Physical Review E* 51 (5): 4282–4286. https://doi.org/10.1103/PhysRevE.51.4282.

Hendrikse, S., J. Treur, and S. Koole. 2023. "Modeling Emerging Interpersonal Synchrony and Its Related Adaptive Short-Term Affiliation and Long-Term Bonding: A Second-Order Multi-Adaptive Neural Agent Model." *International journal of neural systems,* https://doi.org/10.1142/S0129065723500387.

Herrnstein, R. J., and C. Murray. 2010. *The Bell Curve: Intelligence and Class Structure in American Life.* Simon / Schuster.

Heylighen, F. 2009. "Complexity and Self-Organization." In *Encyclopedia of Library and Information Sciences,* 3rd ed., edited by M. J. Bates and M. N. Maack. Boca Raton, FL: CRC Press.

Hirsch, M. W. 1985. "Systems of Differential Equations That Are Competitive or Cooperative II: Convergence Almost Everywhere." *SIAM Journal on Mathematical Analysis* 16 (3): 423–439. https://doi.org/10.1137/0516030.

Hoel, E. P., L. Albantakis, and G. Tononi. 2013. "Quantifying Causal Emergence Shows That Macro Can Beat Micro." *Proceedings of the National Academy of Sciences* 110 (49): 19790–19795. https://doi.org/10.1073/pnas.1314922110.

Hoffmann, G. W., M. W. Benson, G. M. Bree, and P. E. Kinahan. 1986. "A Teachable Neural Network Based on an Unorthodox Neuron." *Physica D: Nonlinear Phenomena,* Proceedings of the Fifth Annual International Conference, 22 (1): 233–246. https://doi.org/10.1016/0167-2789(86)90243-5.

Hofstadter, D. R. 1979. *Gödel, Escher, Bach: An Eternal Golden Braid.* Vintage Books.

———. 2007. *I am a Strange Loop.* New York, NY: Basic Books.

Holland, J. H. 1992a. *Adaptation in Natural and Artificial Systems: An Introductory Analysis with Applications to Biology, Control, and Artificial Intelligence.* MIT Press.

———. 1992b. "Genetic Algorithms." *Scientific American* 267 (1): 66–73. https://doi.org/10.1038/scientificamerican0792-66.

Holyst, J. A., K. Kacperski, and F. Schweitzer. 2001. "Social Impact Models of Opinion Dynamics." In *Annual Reviews of Computational Physics IX,* vol. Volume 9, 253–273. Annual Reviews of Computational Physics. World Scientific. https://doi.org/10.1142/9789812811578_0005.

Huth, K., J. de Ron, J. Luigjes, A. Goudriaan, R. Mohammadi, R. van Holst, E.-J. Wagenmakers, and M. Marsman. 2023. *Bayesian Analysis of Cross-sectional Networks: A Tutorial in R and JASP.* https://doi.org/10.31234/osf.io/ub5tc.

Irvine, E., and M. Sprevak. 2020. "Eliminativism About Consciousness." In *The Oxford Handbook of the Philosophy of Consciousness,* edited by U. Kriegel, 0. Oxford, UK: Oxford University Press. https://doi.org/10.1093/oxfordhb/9780198749677.013.16.

Isvoranu, A.-M., S. Epskamp, L. J. Waldorp, and D. Borsboom. 2022. *Network Psychometrics with R: A Guide for Behavioral and Social Scientists.* Taylor & Francis Limited.

Izhikevich, E. M. 1998. "Multiple Cusp Bifurcations." *Neural Networks* 11 (3): 495–508. https://doi.org/10.1016/S0893-6080(97)00117-2.

Izhikevich, E. M., and R. FitzHugh. 2006. "FitzHugh-Nagumo Model." *Scholarpedia* 1 (9): 1349. https://doi.org/10.4249/scholarpedia.1349.

Jacobson, M. J., J. A. Levin, and M. Kapur. 2019. "Education as a Complex System: Conceptual and Methodological Implications." *Educational Researcher* 48 (2): 112–119. https://doi.org/10.3102/0013189X19826958.

Jager, W. 2017. "Enhancing the Realism of Simulation (EROS): On Implementing and Developing Psychological Theory in Social Simulation." *Journal of Artificial Societies and Social Simulation* 20 (3): 1 14. https://doi.org/10.18564/jasss.3522.

Jager, W., and F. Amblard. 2005. "Uniformity, Bipolarization and Pluriformity Captured as Generic Stylized Behavior with an Agent-Based Simulation Model of Attitude Change." *Computational & Mathematical Organization Theory* 10 (4): 295–303. https://doi.org/10.1007/s10588-005-6282-2.

James, N. A., and D. S. Matteson. 2014. "**Ecp** : An *R* Package for Nonparametric Multiple Change Point Analysis of Multivariate Data." *Journal of Statistical Software* 62 (7). https://doi.org/10.18637/jss.v062.i07.

Jansen, B. R. J., and H. L. J. van der Maas. 2001. "Evidence for the Phase Transition from Rule I to Rule II on the Balance Scale Task." *Developmental Review* 21 (4): 450–494. https://doi.org/10.1006/drev.2001.0530.

Jędrzejewski, A., and K. Sznajd-Weron. 2019. "Statistical Physics of Opinion Formation: Is It a SPOOF?" *Comptes Rendus Physique* 20 (4): 244–261. https://doi.org/10.1016/j.crhy.2019.05.002.

Johnson, C. J., M. A. Mumma, and M.-H. St-Laurent. 2019. "Modeling Multispecies Predator–Prey Dynamics: Predicting the Outcomes of Conservation Actions for Woodland Caribou." *Ecosphere* 10 (3): e02622. https://doi.org/10.1002/ecs2.2622.

Jones, P. J., R. Ma, and R. J. McNally. 2021. "Bridge Centrality: A Network Approach to Understanding Comorbidity." *Multivariate Behavioral Research* 56 (2): 353–367. https://doi.org/10.1080/00273171.2019.1614898.

Jusup, M., P. Holme, K. Kanazawa, M. Takayasu, I. Romić, Z. Wang, S. Geček, et al. 2022. "Social physics." *Physics Reports* 948:1–148. https://doi.org/10.1016/j.physrep.2021.10.005.

Kalantari, S., E. Nazemi, and B. Masoumi. 2020. "Emergence Phenomena in Self-Organizing Systems: A Systematic Literature Review of Concepts, Researches, and Future Prospects." *Journal of Organizational Computing and Electronic Commerce* 30 (3): 224–265. https://doi.org/10.1080/10919392.2020.1748977.

Kalisch, R., A. O. J. Cramer, H. Binder, J. Fritz, I. Leertouwer, G. Lunansk, B. Meyer, J. Timiner, I. M. Veer, and A.-L. Van Harmelen. 2019. "Deconstructing and reconstructing resilience: A dynamic network approach." *Perspectives on Psychological Science* 14 (5): 765–777. https://doi.org/10.1177/1745691619855637.

Karau, S. J., and K. D. Williams. 1993. "Social Loafing: A Meta-Analytic Review and Theoretical Integration." *Journal of Personality and Social Psychology* 65:681–706. https://doi.org/10.1037/0022-3514.65.4.681.

Kargarnovin, S., C. Hernandez, F. V. Farahani, and W. Karwowski. 2023. "Evidence of Chaos in Electroencephalogram Signatures of Human Performance: A Systematic Review." *Brain Sciences* 13 (5): 813. https://doi.org/10.3390/brainsci13050813.

Kauffman, S. A. 1993. *The Origins of Order: Self-Organization and Selection in Evolution.* 1st edition. New York, NY: Oxford University Press.

Kelso, J. A. S. 1995. *Dynamic Patterns: The Self-Organization of Brain and Behavior.* xvii, 334. Cambridge, MA: The MIT Press.

———. 2021. "The Haken–Kelso–Bunz (HKB) Model: From Matter to Movement to Mind." *Biological Cybernetics* 115 (4): 305–322. https://doi.org/10.1007/s00422-021-00890-w.

Kelso, J. A. S., J. P. Scholz, and G. Schöner. 1986. "Nonequilibrium Phase Transitions in Coordinated Biological Motion: Critical Fluctuations." *Physics Letters A* 118 (6): 279–284. https://doi.org/10.1016/0375-9601(86)90359-2.

Kelso, J. A. S., and G. Schöner. 1988. "Self-Organization of Coordinative Movement Patterns." *Human Movement Science* 7 (1): 27–46. https://doi.org/10.1016/0167-9457(88)90003-6.

Kim, J. 2006. "Emergence: Core Ideas and Issues." *Synthese* 151 (3): 547–559. https://doi.org/10.1007/s11229-006-9025-0.

Klinkenberg, S., M. Straatemeier, and H. L. J. van der Maas. 2011. "Computer Adaptive Practice of Maths Ability Using a New Item Response Model for on the Fly Ability and Difficulty Estimation." *Computers & Education* 57:1813–1824.

Klose, A. K., V. Karle, R. Winkelmann, and J. F. Donges. 2020. "Emergence of Cascading Dynamics in Interacting Tipping Elements of Ecology and Climate." *Royal Society Open Science* 7 (6): 200599. https://doi.org/10.1098/rsos.200599.

Knyspel, J., and R. Plomin. 2024. "Comparing Factor and Network Models of Cognitive Abilities Using Twin Data." *Intelligence* 104:101833. https://doi.org/10.1016/j.intell.2024.101833.

Kossakowski, J. J., L. J. Waldorp, and H. L. J. van der Maas. 2021. "The Search for Causality: A Comparison of Different Techniques for Causal Inference Graphs." *Psychological Methods* 26:719–742. https://doi.org/10.1037/met0000390.

Kosterlitz, J. M. 1974. "The Critical Properties of the Two-Dimensional Xy Model." *Journal of Physics C: Solid State Physics* 7 (6): 1046. https://doi.org/10.1088/0022-3719/7/6/005.

Krakauer, D. C. 2023. "Symmetry–Simplicity, Broken Symmetry–Complexity." *Interface Focus* 13 (3): 20220075. https://doi.org/10.1098/rsfs.2022.0075.

———. 2024. "The Complex World: An Introduction to the Foundations of Complexity Science." In *Foundational Papers in Complexity Science,* edited by D. C. Krakauer, vol. 1, Intro 1–99. Santa Fe, NM: SFI Press.

Kreitler, S., and H. Kreitler. 1989. "Horizontal Decalage: A Problem and Its Solution." *Cognitive Development* 4 (1): 89–119. https://doi.org/10.1016/0885-2014(89)90006-3.

Kruis, J., G. K. J. Maris, M. Marsman, M. Bolsinova, and H. L. J. van der Maas. 2020. "Deviations of Rational Choice: An Integrative Explanation of the Endowment and Several Context Effects." *Scientific Reports* 10 (1): 16226. https://doi.org/10.1038/s41598-020-73181-2.

Kruse, P., and M. Stadler. 2012. *Ambiguity in Mind and Nature: Multistable Cognitive Phenomena.* Springer Science & Business Media.

Kumar, A. A. 2021. "Semantic Memory: A Review of Methods, Models, and Current Challenges." *Psychonomic Bulletin & Review* 28 (1): 40–80. https://doi.org/10.3758/s13423-020-01792-x.

Kuramoto, Y. 1984. "Chemical Turbulence." In *Chemical Oscillations, Waves, and Turbulence,* edited by Y. Kuramoto, 111–140. Springer Series in Synergetics. Berlin, Heidelberg. https://doi.org/10.1007/978-3-642-69689-3_7.

Ladyman, J., J. Lambert, and K. Wiesner. 2013. "What Is a Complex System?" *European Journal for Philosophy of Science* 3 (1): 33–67. https://doi.org/10.1007/s13194-012-0056-8.

Lange, J., and J. H. Zickfeld. 2021. "Emotions as Overlapping Causal Networks of Emotion Components: Implications and Methodological Approaches." *Emotion Review* 13 (2): 157–167. https://doi.org/10.1177/1754073920988787.

Langton, C. G. 1990. "Computation at the Edge of Chaos: Phase Transitions and Emergent Computation." *Physica D: Nonlinear Phenomena* 42 (1): 12–37. https://doi.org/10.1016/0167-2789(90)90064-V.

Latané, B. 1981. "The Psychology of Social Impact." *American Psychologist* 36:343–356. https://doi.org/10.1037/0003-066X.36.4.343.

Latané, B., and A. Nowak. 1994. "Attitudes as Catastrophes: From Dimensions to Categories with Increasing Involvement." In *Dynamical Systems in Social Psychology,* 219–249. San Diego, CA: Academic Press.

Leggio, L., G. A. Kenna, M. Fenton, E. Bonenfant, and R. M. Swift. 2009. "Typologies of Alcohol Dependence. From Jellinek to Genetics and Beyond." *Neuropsychology Review* 19 (1): 115–129. https://doi.org/10.1007/s11065-008-9080-z.

Lei, M., Z. Wang, and Z. Feng. 2001. "Detecting Nonlinearity of Action Surface EMG Signal." *Physics Letters A* 290 (5): 297–303. https://doi.org/10.1016/S0375-9601(01)00668-5.

Lemke, J. L., and N. H. Sabelli. 2008. "Complex Systems and Educational Change: Towards a New Research Agenda." Chap. 8 in *Complexity Theory and the Philosophy of Education,* edited by M. Mason, 112–123. https://doi.org/10.1002/9781444307351.ch8.

Lewenstein, M., A. Nowak, and B. Latané. 1992. "Statistical Mechanics of Social Impact." *Physical Review A* 45 (2): 763–776. https://doi.org/10.1103/PhysRevA.45.763.

Li, H., K. Liang, G. Kou, and Y. Dong. 2020. "Opinion Dynamics Model Based on the Cognitive Dissonance: An Agent-Based Simulation." *Information Fusion* 56:1–14. https://doi.org/10.1016/j.inffus.2019.09.006.

Loehle, C. 1989. "Catastrophe Theory in Ecology: A Critical Review and an Example of the Butterfly Catastrophe." *Ecological Modelling* 49 (1): 125–152. https://doi.org/10.1016/0304-3800(89)90047-1.

Lorenz, E. N. 1963. "Deterministic Nonperiodic Flow." *Journal of the Atmospheric Sciences* 20 (2): 130–141. https://doi.org/10.1175/1520-0469(1963)020<0130:DNF>2.0.CO;2.

Lorenz, J. 2007. "Continuous Opinion Dynamics under Bounded Confidence: A Survey." *International Journal of Modern Physics C* 18 (12): 1819–1838. https://doi.org/10.1142/S0129183107011789.

Love, J., R. Selker, M. Marsman, T. Jamil, D. Dropmann, J. Verhagen, A. Ly, et al. 2019. "JASP: Graphical Statistical Software for Common Statistical Designs." *Journal of Statistical Software* 88. https://doi.org/10.18637/jss.v088.i02.

Lowery, M. R., M. A. Clark, and N. T. Carter. 2021. "The Balancing Act of Performance: Psychometric Networks and the Causal Interplay of Organizational Citizenship and Counterproductive Work Behaviors." *Journal of Vocational Behavior* 125:103527. https://doi.org/10.1016/j.jvb.2020.103527.

Ludwig, D., D. D. Jones, and C. S. Holling. 1978. "Qualitative Analysis of Insect Outbreak Systems: The Spruce Budworm and Forest." *Journal of Animal Ecology* 47 (1): 315–332. https://doi.org/10.2307/3939.

Lumsden, J. 1976. "Test Theory." *Annual Review of Psychology* 27:251–280. https://doi.org/10.1146/annurev.ps.27.020176.001343.

Lunansky, G., J. Naberman, C. D. van Borkulo, C. Chen, L. Wang, and D. Borsboom. 2022. "Intervening on Psychopathology Networks: Evaluating Intervention Targets through Simulations." *Methods* 204:29–37. https://doi.org/10.1016/j.ymeth.2021.11.006.

Luppi, A. I., P. A. M. Mediano, F. E. Rosas, N. Holland, T. D. Fryer, J. T. O'Brien, J. B. Rowe, D. K. Menon, D. Bor, and E. A. Stamatakis. 2022. "A Synergistic Core for Human Brain Evolution and Cognition." *Nature Neuroscience* 25 (6): 771–782. https://doi.org/10.1038/s41593-022-01070-0.

Lurie, D. J., D. Kessler, D. S. Bassett, R. F. Betzel, M. Breakspear, S. Kheilholz, A. Kucyi, et al. 2020. "Questions and Controversies in the Study of Time-Varying Functional Connectivity in Resting fMRI." *Network Neuroscience* 4 (1): 30–69. https://doi.org/10.1162/netn_a_00116.

Macy, M. W., A. Flache, and K. Takacs. 2006. "What Sustains Stable Cultural Diversity and What Undermines It? Axelrod and beyond. Peer Review." In *Advancing Social Simulation: Proceedings of the First World Congress on Social Simulation,* edited by S. Takahashi, 9–16. Kyoto, Japan: Springer.

Mahmoud, H. 2008. *Polya Urn Models.* Boca Raton, FL: CRC Press.

Maris, G. K. J., and H. L. J. van der Maas. 2012. "Speed-Accuracy Response Models: Scoring Rules Based on Response Time and Accuracy." *Psychometrika* 77 (4): 615–633. https://doi.org/10.1007/s11336-012-9288-y.

Marsman, M., D. Borsboom, J. Kruis, S. Epskamp, R. van Bork, L. J. Waldorp, H. L. J. van der Maas, and G. K. J. Maris. 2018. "An Introduction to Network Psychometrics: Relating Ising Network Models to Item Response Theory Models." *Multivariate Behavioral Research* 53 (1): 15–35. https://doi.org/10.1080/00273171.2017.1379379.

Marsman, M., and M. Rhemtulla. 2022. "Guest Editors' Introduction to The Special Issue "Network Psychometrics in Action": Methodological Innovations Inspired by Empirical Problems." *Psychometrika* 87 (1): 1–11. https://doi.org/10.1007/s11336-022-09861-x.

Martins, A. C. R. 2008. "Continuous Opinions and Discrete Actions in Opinion Dynamics Problems." *International Journal of Modern Physics C* 19 (04): 617–624. https://doi.org/10.1142/S0129183108012339.

Masuda, N. 2014. "Voter Model on the Two-Clique Graph." *Physical Review E* 90 (1): 012802. https://doi.org/10.1103/PhysRevE.90.012802.

Mazanov, J., and D. G. Byrne. 2006. "A Cusp Catastrophe Model Analysis of Changes in Adolescent Substance Use: Assessment of Behavioural Intention as a Bifurcation Variable." *Nonlinear Dynamics, Psychology, and Life Sciences* 10 (4): 445–470.

McClelland, J. L., and D. E. Rumelhart. 1988. *Explorations in Parallel Distributed Processing: A Handbook of Models, Programs, and Exercises.* ix, 344. Cambridge, MA: The MIT Press.

McGeer, T. 1990. "Passive Walking with Knees." In *Proceedings., IEEE International Conference on Robotics and Automation Proceedings,* 3:1640–1645. https://doi.org/10.1109/ROBOT.1990.126245.

McGrath, T., A. Kapishnikov, N. Tomašev, A. Pearce, M. Wattenberg, D. Hassabis, B. Kim, U. Paquet, and V. Kramnik. 2022. "Acquisition of Chess Knowledge in AlphaZero." *Proceedings of the National Academy of Sciences* 119 (47): e2206625119. https://doi.org/10.1073/pnas.2206625119.

McLachlan, G. J., S. X. Lee, and S. I. Rathnayake. 2019. "Finite Mixture Models." *Annual Review of Statistics and Its Application* 6 (1): 355–378. https://doi.org/10.1146/annurev-statistics-031017-100325.

McRobie, A. 2017. *The Seduction of Curves: The Lines of Beauty That Connect Mathematics, Art, and the Nude.* Princeton, NJ: Princeton University Press.

Meadows, D. H. 2008. *Thinking in Systems: A Primer.* Chelsea Green Publishing.

Meena, C., C. Hens, S. Acharyya, S. Haber, S. Boccaletti, and B. Barzel. 2023. "Emergent Stability in Complex Network Dynamics." *Nature Physics* 19:1–10. https://doi.org/10.1038/s41567-023-02020-8.

Meier, S. T. 2013. *The Chronic Crisis in Psychological Measurement and Assessment: A Historical Survey.* San Diego, CA: Academic Press.

Michell, J. 1997. "Quantitative Science and the Definition of Measurement in Psychology." *British Journal of Psychology* 88 (3): 355–383. https://doi.org/10.1111/j.2044-8295.1997.tb02641.x.

———. 1999. *Measurement in Psychology: A Critical History of a Methodological Concept.* New York, NY: Cambridge University Press.

———. 2008. "Is Psychometrics Pathological Science?" *Measurement: Interdisciplinary Research and Perspectives* 6:7–24. https://doi.org/10.1080/15366360802035489.

Miller, J. H., and S. E. Page. 2007. *Complex Adaptive Systems: An Introduction to Computational Models of Social Life.* xix, 263. Princeton, NJ: Princeton University Press.

Mitchell, M. 1996. *An Introduction to Genetic Algorithms.* Cambridge, UK: MIT Press.

———. 2009. *Complexity: A Guided Tour.* Oxford, UK: Oxford University Press.

Molenaar, P. C. M. 2004. "A Manifesto on Psychology as Idiographic Science: Bringing the Person Back Into Scientific Psychology, This Time Forever." *Measurement: Interdisciplinary Research and Perspectives* 2 (4): 201–218. https://doi.org/10.1207/s15366359mea0204_1.

Molenaar, P. C. M., D. I. Boomsma, and C. V. Dolan. 1993. "A Third Source of Developmental Differences." *Behavior Genetics* 23 (6): 519–524. https://doi.org/10.1007/BF01068142.

Molenaar, P. C. M., and K. M. Newell. 2003. "Direct Fit of a Theoretical Model of Phase Transition in Oscillatory Finger Motions." *British Journal of Mathematical and Statistical Psychology* 56 (2): 199–214. https://doi.org/10.1348/000711003770480002.

Molenaar, P. C. M., and L. Oppenheimer. 1985. "Dynamic Models of Development and the Mechanistic-Organismic Controversy." *New Ideas in Psychology* 3 (3): 233–242. https://doi.org/10.1016/0732-118X(85)90017-0.

Monroe, B. M., and S. J. Read. 2008. "A General Connectionist Model of Attitude Structure and Change: The ACS (Attitudes as Constraint Satisfaction) Model." *Psychological Review* 115:733–759. https://doi.org/10.1037/0033-295X.115.3.733.

Monsivais-Velazquez, D., K. Bhattacharya, R. A. Barrio, P. K. Maini, and K. K. Kaski. 2020. "Dynamics of Hierarchical Weighted Networks of van Der Pol Oscillators." *Chaos (Woodbury, N.Y.)* 30 (12): 123146. https://doi.org/10.1063/5.0010638.

Morel, B., and R. Ramanujam. 1999. "Through the Looking Glass of Complexity: The Dynamics of Organizations as Adaptive and Evolving Systems." *Organization Science* 10 (3): 278–293. https://doi.org/10.1287/orsc.10.3.278.

Murray, J. D. 1989. *Mathematical Biology.* Berlin, Heidelberg: Springer. https://doi.org/10.1007/978-3-662-08539-4.

Nash, J. F. 1950. "Equilibrium Points in N-Person Games." *Proceedings of the National Academy of Sciences* 36 (1): 48–49. https://doi.org/10.1073/pnas.36.1.48.

Needham, A. W., and E. L. Nelson. 2023. "How Babies Use Their Hands to Learn about Objects: Exploration, Reach-to-Grasp, Manipulation, and Tool Use." *WIREs Cognitive Science* n/a (n/a): e1661. https://doi.org/10.1002/wcs.1661.

Nenu, T. 2022. "Douglas Hofstadter's Gödelian Philosophy of Mind." *Journal of Artificial Intelligence and Consciousness* 09 (02): 241–266. https://doi.org/10.1142/S2705078522500011.

Newman, M. E. J., and J. Park. 2003. "Why Social Networks Are Different from Other Types of Networks." *Physical Review E* 68 (3): 036122. https://doi.org/10.1103/PhysRevE.68.036122.

Nisbett, R. E., J. Aronson, C. Blair, W. Dickens, J. Flynn, D. F. Halpern, and E. Turkheimer. 2012. "Intelligence: New Findings and Theoretical Developments." *American Psychologist* 67 (2): 130–159. https://doi.org/10.1037/a0026699.

Noorazar, H. 2020. "Recent Advances in Opinion Propagation Dynamics: A 2020 Survey." *The European Physical Journal Plus* 135 (6): 521. https://doi.org/10.1140/epjp/s13360-020-00541-2.

Nosek, B. A., T. E. Hardwicke, H. Moshontz, A. Allard, K. S. Corker, A. Dreber, F. Fidler, et al. 2022. "Replicability, Robustness, and Reproducibility in Psychological Science." *Annual Review of Psychology* 73 (1): 719–748. https://doi.org/10.1146/annurev-psych-020821-114157.

Nowak, A., J. Szamrej, and L. Bibb. 1990. "From Private Attitude to Public Opinion: A Dynamic Theory of Social Impact." *Psychological Review* 97 (3): 362–76. https://doi.org/10.1037/0033-295X.97.3.362.

O'Byrne, J., and K. Jerbi. 2022. "How Critical Is Brain Criticality?" *Trends in Neurosciences* 45 (11): 820–837. https://doi.org/10.1016/j.tins.2022.08.007.

Oberauer, K., and S. Lewandowsky. 2019. "Addressing the Theory Crisis in Psychology." *Psychonomic Bulletin & Review* 26 (5): 1596–1618. https://doi.org/10.3758/s13423-019-01645-2.

Oliva, T. A., W. S. Desarbo, D. L. Day, and K. Jedidi. 1987. "Gemcat: A General Multivariate Methodology for Estimating Catastrophe Models." *Behavioral Science* 32 (2): 121–137. https://doi.org/10.1002/bs.3830320205.

Olthof, M., F. Hasselman, F. Oude Maatman, A. M. T. Bosman, and A. Lichtwarck-Aschoff. 2023. "Complexity Theory of Psychopathology." *Journal of Psychopathology and Clinical Science* 132:314–323. https://doi.org/10.1037/abn0000740.

Olthof, M., F. Hasselman, G. Strunk, M. van Rooij, B. Aas, M. A. Helmich, G. Schiepek, and A. Lichtwarck-Aschoff. 2020. "Critical Fluctuations as an Early-Warning Signal for Sudden Gains and Losses in Patients Receiving Psychotherapy for Mood Disorders." *Clinical Psychological Science* 8 (1): 25–35. https://doi.org/10.1177/2167702619865969.

Parisi, G. 2023. *In a Flight of Starlings: The Wonders of Complex Systems.* New York, NY: Penguin Publishing Group.

Pashler, H., and E.-J. Wagenmakers. 2012. "Editors' Introduction to the Special Section on Replicability in Psychological Science: A Crisis of Confidence?" *Perspectives on Psychological Science* 7:528–530. https://doi.org/10.1177/1745691612465253.

Paulos, J. A. 2008. *Mathematics and Humor.* Chicago, IL: University of Chicago Press.

Pavlus, J. 2016. "The Clumsy Quest to Perfect the Walking Robot." *Scientific American,* https://doi.org/10.1038/scientificamerican0716-60.

Penke, L., J. J. A. Denissen, and G. F. Miller. 2007. "The Evolutionary Genetics of Personality." *European Journal of Personality* 21 (5): 549–587. https://doi.org/10.1002/per.629.

Peralta, A. F., J. Kertész, and G. Iñiguez. 2022. *Opinion Dynamics in Social Networks: From Models to Data.* https://doi.org/10.48550/arXiv.2201.01322.

Perc, M., J. J. Jordan, D. G. Rand, Z. Wang, S. Boccaletti, and A. Szolnoki. 2017. "Statistical Physics of Human Cooperation." *Physics Reports* 687:1–51. https://doi.org/10.1016/j.physrep.2017.05.004.

Peters, G. J. Y., S. Zörgő, and H. L. J. van der Maas. 2022. *The Qualitative Network Approach (QNA.* https://doi.org/10.31234/osf.io/cvf52.

Piaget, J. 1952. *The Origins of Intelligence in Children.* Edited by M. Cook. 419. New York, NY. https://doi.org/10.1037/11494-000.

Plenz, D., T. L. Ribeiro, S. R. Miller, P. A. Kells, A. Vakili, and E. L. Capek. 2021. "Self-Organized Criticality in the Brain." *Frontiers in Physics* 9. https://doi.org/10.3389/fphy.2021.639389.

Ploeger, A., H. L. J. van der Maas, and P. A. I. Hartelman. 2002. "Stochastic Catastrophe Analysis of Switches in the Perception of Apparent Motion." *Psychonomic Bulletin & Review* 9 (1): 26–42. https://doi.org/10.3758/BF03196255.

Pool, R. 1989. "Is It Healthy to Be Chaotic?" *Science* 243 (4891): 604–607. https://doi.org/10.1126/science.2916117.

Port, R. F., and T. Van Gelder. 1995. *Mind as Motion: Explorations in the Dynamics of Cognition.* Cambridge, MA: MIT Press.

Posternak, M. A., and I. Miller. 2001. "Untreated Short-Term Course of Major Depression: A Meta-Analysis of Outcomes from Studies Using Wait-List Control Groups." *Journal of Affective Disorders* 66 (2): 139–146. https://doi.org/10.1016/S0165-0327(00)00304-9.

Poston, T., and I. Stewart. 1978. *Catastrophe Theory and Its Applications.* London, UK: Pitman.

Pritchard, W. S., and D. W. Duke. 1992. "Measuring Chaos in the Brain: A Tutorial Review of Nonlinear Dynamical Eeg Analysis." *International Journal of Neuroscience* 67 (1-4): 31–80. https://doi.org/10.3109/00207459208994774.

Proskurnikov, A. V., and R. Tempo. 2017. "A Tutorial on Modeling and Analysis of Dynamic Social Networks. Part I." *Annual Reviews in Control* 43:65–79. https://doi.org/10.1016/j.arcontrol.2017.03.002.

———. 2018. "A Tutorial on Modeling and Analysis of Dynamic Social Networks. Part II." *Annual Reviews in Control* 45:166–190. https://doi.org/10.1016/j.arcontrol.2018.03.005.

Rapoport, A. 1960. *Fights, Games, and Debates.* University of Michigan Press.

Ratcliff, R., P. L. Smith, S. D. Brown, and G. McKoon. 2016. "Diffusion Decision Model: Current Issues and History." *Trends in Cognitive Sciences* 20 (4): 260–281. https://doi.org/10.1016/j.tics.2016.01.007.

Read, J., and J. Moncrieff. 2022. "Depression: Why Drugs and Electricity Are Not the Answer." *Psychological Medicine* 52 (8): 1401–1410. https://doi.org/10.1017/S0033291721005031.

Redner, S. 2019. "Reality-Inspired Voter Models: A Mini-Review." *Comptes Rendus Physique* 20 (4): 275–292. https://doi.org/10.1016/j.crhy.2019.05.004.

Reher, J., and A. D. Ames. 2021. "Dynamic Walking: Toward Agile and Efficient Bipedal Robots." *Annual Review of Control, Robotics, and Autonomous Systems* 4 (1): 535–572. https://doi.org/10.1146/annurev-control-071020-045021.

Remmerswaal, D., J. Jongerling, P. J. Jansen, C. Eielts, and I. H. A. Franken. 2019. "Impaired Subjective Self-Control in Alcohol Use: An Ecological Momentary Assessment Study." *Drug and Alcohol Dependence* 204:107479. https://doi.org/10.1016/j.drugalcdep.2019.04.043.

Rendell, P. 2016. "Game of Life Universal Turing Machine." In *Turing Machine Universality of the Game of Life,* edited by P. Rendell, 71–89. Emergence, Complexity and Computation. Cham: Springer International Publishing. https://doi.org/10.1007/978-3-319-19842-2_5.

Repp, B. H., and Y.-H. Su. 2013. "Sensorimotor Synchronization: A Review of Recent Research (2006–2012)." *Psychonomic Bulletin & Review* 20 (3): 403–452. https://doi.org/10.3758/s13423-012-0371-2.

Rinaldi, S. 1998. "Love Dynamics: The Case of Linear Couples." *Applied Mathematics and Computation* 95 (2): 181–192. https://doi.org/10.1016/S0096-3003(97)10081-9.

Roberts, J. A., L. L. Gollo, R. G. Abeysuriya, G. Roberts, P. B. Mitchell, M. W. Woolrich, and M. Breakspear. 2019. "Metastable Brain Waves." *Nature Communications* 10 (1): 1056. https://doi.org/10.1038/s41467-019-08999-0.

Robertson, R., and A. Combs, eds. 2014. *Chaos Theory in Psychology and the Life Sciences.* New York, NY: Psychology Press. https://doi.org/10.4324/9781315806280.

Robinaugh, D. J., J. Haslbeck, L. J. Waldorp, J. Kossakowski, E. I. Fried, A. Millner, R. J. McNally, et al. 2019. *Advancing the Network Theory of Mental Disorders: A Computational Model of Panic Disorder.* https://doi.org/10.31234/osf.io/km37w.

Robinaugh, D. J., R. H. A. Hoekstra, E. R. Toner, and D. Borsboom. 2020. "The Network Approach to Psychopathology: A Review of the Literature 2008–2018 and an Agenda for Future Research." *Psychological Medicine* 50 (3): 353–366. https://doi.org/10.1017/S0033291719003404.

Rodríguez-Bermúdez, G., and P. J. García-Laencina. 2015. "Analysis of EEG Signals using Nonlinear Dynamics and Chaos: A Review." *Applied Mathematics & Information Science* 9 (5): 2309–2321. https://www.naturalspublishing.com/files/published/d4c96j5fp99493.pdf.

Rosencrans, P. L., L. A. Zoellner, and N. C. Feeny. 2021. "A Network Approach to Posttraumatic Stress Disorder: Comparing Interview and Self-Report Networks." *Psychological Trauma: Theory, Research, Practice, and Policy,* No Pagination Specified–No Pagination Specified. https://doi.org/10.1037/tra0001151.

Rosser, J. B. 2007. "The Rise and Fall of Catastrophe Theory Applications in Economics: Was the Baby Thrown out with the Bathwater?" *Journal of Economic Dynamics and Control* 31 (10): 3255–3280. https://doi.org/10.1016/j.jedc.2006.09.013.

Rosso, O. A., H. A. Larrondo, M. T. Martin, A. Plastino, and M. A. Fuentes. 2007. "Distinguishing Noise from Chaos." *Physical Review Letters* 99 (15): 154102. https://doi.org/10.1103/PhysRevLett.99.154102.

Roxin, A., and A. Ledberg. 2008. "Neurobiological Models of Two-Choice Decision Making Can Be Reduced to a One-Dimensional Nonlinear Diffusion Equation." *PLOS Computational Biology* 4 (3): e1000046. https://doi.org/10.1371/journal.pcbi.1000046.

Rutjens, B. T., S. J. Heine, R. M. Sutton, and F. van Harreveld. 2018. "Chapter Three - Attitudes Towards Science." In *Advances in Experimental Social Psychology,* edited by J. M. Olson, 57:125–165. Academic Press. https://doi.org/10.1016/bs.aesp.2017.08.001.

Ryan, O., L. F. Bringmann, and N. K. Schuurman. 2022. "The Challenge of Generating Causal Hypotheses Using Network Models." *Structural Equation Modeling: A Multidisciplinary Journal* 29 (6): 953–970. https://doi.org/10.1080/10705511.2022.2056039.

Sachisthal, M. S. M., B. R. J. Jansen, T. T. D. Peetsma, J. Dalege, H. L. J. van der Maas, and M. E. J. Raijmakers. 2019. "Introducing a Science Interest Network Model to Reveal Country Differences." *Journal of Educational Psychology* 111:1063–1080. https://doi.org/10.1037/edu0000327.

Sala, G., N. D. Aksayli, K. S. Tatlidil, T. Tatsumi, Y. Gondo, and F. Gobet. 2019. "Near and Far Transfer in Cognitive Training: A Second-Order Meta-Analysis." Edited by R. Zwaan and P. Verkoeijen. *Collabra: Psychology* 5 (1): 18. https://doi.org/10.1525/collabra.203.

Salanti, G., C. Del Giovane, A. Chaimani, D. M. Caldwell, and J. P. T. Higgins. 2014. "Evaluating the Quality of Evidence from a Network Meta-Analysis." *PLoS One* 9 (7): e99682. https://doi.org/10.1371/journal.pone.0099682.

Sandubete, J. E., and L. Escot. 2021. "DChaos: An R Package for Chaotic Time Series Analysis." *The R Journal* 13 (1): 232. https://doi.org/10.32614/RJ-2021-036.

Saul, D. M., M. Wortis, and D. Stauffer. 1974. "Tricritical Behavior of the Blume-Capel Model." *Physical Review B* 9 (11): 4964–4980. https://doi.org/10.1103/PhysRevB.9.4964.

Savi, A. O., M. Marsman, H. L. J. van der Maas, and G. K. J. Maris. 2019. "The Wiring of Intelligence." *Perspectives on Psychological Science* 14 (6): 1034–1061. https://doi.org/10.1177/1745691619866447.

Sayama, H. 2015. "Introduction to the Modeling and Analysis of Complex Systems." *Milne Open Textbooks,* https://knightscholar.geneseo.edu/oer-ost/14.

Scheffer, M. 2004. *Ecology of Shallow Lakes.* Dordrecht. https://doi.org/10.1007/978-1-4020-3154-0.

Scheffer, M., C. L. Bockting, D. Borsboom, R. Cools, C. Delecroix, J. A. Hartmann, K. S. Kendler, et al. 2024. "A Dynamical Systems View of Psychiatric Disorders—Practical Implications: A Review." *JAMA Psychiatry,* https://doi.org/10.1001/jamapsychiatry.2024.0228.

Schelling, T. C. 1971. "Dynamic Models of Segregation." *The Journal of Mathematical Sociology* 1 (2): 143–186. https://doi.org/10.1080/0022250X.1971.9989794.

Schiepek, G., and V. Perlitz. 2009. "Self-Organization in Clinical Psychology." In *Synergetics,* edited by A. Hutt and H. Haken, 263–285. New York, NY: Springer. https://doi.org/10.1007/978-1-0716-0421-2_472.

Schiepek, G. K., K. Viol, W. Aichhorn, M.-T. Hütt, K. Sungler, D. Pincus, and H. J. Schöller. 2017. "Psychotherapy Is Chaotic—(Not Only) in a Computational World." *Frontiers in Psychology* 8. https://www.frontiersin.org/articles/10.3389/fpsyg.2017.00379.

Schmidhuber, J. 2015. "Deep Learning in Neural Networks: An Overview." *Neural Networks* 61:85–117. https://doi.org/10.1016/j.neunet.2014.09.003.

Schmidt, R. C., C. Carello, and M. T. Turvey. 1990. "Phase Transitions and Critical Fluctuations in the Visual Coordination of Rhythmic Movements between People." *Journal of Experimental Psychology. Human Perception and Performance* 16 (2): 227–247. https://doi.org/10.1037//0096-1523.16.2.227.

Schneider, W. J., and K. S. McGrew. 2012. "The Cattell–Horn–Carroll Model of Intelligence." In *Contemporary Intellectual Assessment: Theories, Tests, and Issues, 3rd Ed,* 99–144. New York, NY: The Guilford Press.

Schöner, G. 2020. "The Dynamics of Neural Populations Capture the Laws of the Mind." *Topics in Cognitive Science* 12 (4): 1257–1271. https://doi.org/10.1111/tops.12453.

Schöner, G., and J. P. Spencer. 2016. *Dynamic Thinking: A Primer on Dynamic Field Theory.* New York, NY: Oxford University Press.

Schwartz, S. J., S. O. Lilienfeld, A. Meca, and K. C. Sauvigné. 2016. "The role of neuroscience within psychology: A call for inclusiveness over exclusiveness." *American Psychologist* 71 (1): 52–70. https://doi.org/10.1037/a0039678.

Schweren, L., C. D. van Borkulo, E. I. Fried, and I. M. Goodyer. 2018. "Assessment of Symptom Network Density as a Prognostic Marker of Treatment Response in Adolescent Depression." *JAMA Psychiatry* 75 (1): 98–100. https://doi.org/10.1001/jamapsychiatry.2017.3561.

Scott, J. 2011. "Social Network Analysis: Developments, Advances, and Prospects." *Social Network Analysis and Mining* 1 (1): 21–26. https://doi.org/10.1007/s13278-010-0012-6.

Sen, P., and B. K. Chakrabarti. 2014. *Sociophysics: An Introduction.* OUP Oxford.

Serra, R., and G. Zanarini. 1990. *Complex Systems and Cognitive Processes.* Berlin, Heidelberg: Springer. https://doi.org/10.1007/978-3-642-46678-6.

Seth, A. K., and T. Bayne. 2022. "Theories of consciousness." *Nature Reviews Neuroscience* 23 (7): 439–452. https://doi.org/10.1038/s41583-022-00587-4.

Sherif, M., and C. I. Hovland. 1961. *Social Judgment: Assimilation and Contrast Effects in Communication and Attitude Change.* xii, 218. New Haven, CT: Yale University Press.

Silver, D., T. Hubert, J. Schrittwieser, I. Antonoglou, M. Lai, A. Guez, M. Lanctot, et al. 2018. "A General Reinforcement Learning Algorithm That Masters Chess, Shogi, and Go through Self-Play." *Science* 362 (6419): 1140–1144. https://doi.org/10.1126/science.aar6404.

Simon, H. A. 1962. "The Architecture of Complexity." *Proceedings of the American Philosophical Society* 106 (6): 467–482. https://www.jstor.org/stable/985254.

Skarda, C. A., and W. J. Freeman. 1987. "How Brains Make Chaos in Order to Make Sense of the World." *Behavioral and Brain Sciences* 10 (2): 161–173. https://doi.org/10.1017/S0140525X00047336.

Smal, I., J. Dalege, and H. L. J. van der Maas. Submitted. "The Learning Ising Attitude Model (LIAM): Entropy Reduction by Hebbian Learning."

Smaldino, P. 2023. *Modeling Social Behavior: Mathematical and Agent-Based Models of Social Dynamics and Cultural Evolution.* Princeton University Press.

Smolensky, P. 1986. *Information Processing in Dynamical Systems: Foundations of Harmony Theory.* Technical report.

Snijders, T. A. B. 2001. "The Statistical Evaluation of Social Network Dynamics." *Sociological Methodology* 31 (1): 361–395. https://doi.org/10.1111/0081-1750.00099.

Sobkowicz, P. 2012. "Discrete Model of Opinion Changes Using Knowledge and Emotions as Control Variables." *PLoS ONE* 7 (9): e44489. https://doi.org/10.1371/journal.pone.0044489.

———. 2013. "Minority Persistence in Agent Based Model Using Information and Emotional Arousal as Control Variables." *The European Physical Journal B* 86 (7): 335. https://doi.org/10.1140/epjb/e2013-40029-0.

———. 2020. "Whither Now, Opinion Modelers?" *Frontiers in Physics* 8. https://doi.org/10.3389/fphy.2020.587009.

Soetaert, K., T. Petzoldt, and R. W. Setzer. 2010. "Solving Differential Equations in *R*: Package deSolve." *Journal of Statistical Software* 33 (9). https://doi.org/10.18637/jss.v033.i09.

Spearman, C. 1904. "'General Intelligence,' Objectively Determined and Measured." *The American Journal of Psychology* 15:201–293. https://doi.org/10.2307/1412107.

Spiller, T. R., O. Levi, Y. Neria, B. Suarez-Jimenez, Y. Bar-Haim, and A. Lazarov. 2020. "On the Validity of the Centrality Hypothesis in Cross-Sectional between-Subject Networks of Psychopathology." *BMC Medicine* 18 (1): 297. https://doi.org/10.1186/s12916-020-01740-5.

Sprott, J. C. 2004. "Dynamical Models of Love." *Nonlinear Dynamics, Psychology, and Life Sciences* 8 (3): 303–314.

———. 2005. "Dynamical Models of Happiness." *Nonlinear Dynamics, Psychology, and Life Sciences* 9:23–36.

Stam, C. J. 2005. "Nonlinear Dynamical Analysis of EEG and MEG: Review of an Emerging Field." *Clinical Neurophysiology: Official Journal of the International Federation of Clinical Neurophysiology* 116 (10): 2266–2301. https://doi.org/10.1016/j.clinph.2005.06.011.

Steels, L. 1995. "A Self-Organizing Spatial Vocabulary." *Artificial Life* 2 (3): 319–332. https://doi.org/10.1162/artl.1995.2.3.319.

Stefan, M. 2020. *RPubs - Fractals with R.* https://rpubs.com/mstefan-rpubs/fractals.

Stengers, I., and I. Prigogine. 1984. *Order Out of Chaos: Man's New Dialogue with Nature.* Toronto: Bantam Books.

Stepp, N., and T. D. Frank. 2009. "A Data-Analysis Method for Decomposing Synchronization Variability of Anticipatory Systems into Stochastic and Deterministic Components." *The European Physical Journal B* 67 (2): 251–257. https://doi.org/10.1140/epjb/e2009-00022-x.

Stevens, S. S. 1957. "On the Psychophysical Law." *Psychological Review* 64 (3): 153–181. https://doi.org/10.1037/h0046162.

Stewart, I. 1982. "Catastrophe Theory in Physics." *Reports on Progress in Physics* 45 (2): 185–221. https://doi.org/10.1088/0034-4885/45/2/002.

Stewart, Ian, and P. L. Peregoy. 1983. "Catastrophe Theory Modeling in Psychology." *Psychological Bulletin* 94:336–362. https://doi.org/10.1037/0033-2909.94.2.336.

Steyvers, M., and J. B. Tenenbaum. 2005. "The Large-Scale Structure of Semantic Networks: Statistical Analyses and a Model of Semantic Growth." *Cognitive Science* 29 (1): 41–78. https://doi.org/10.1207/s15516709cog2901_3.

Stouffer, S. A., E. A. Suchman, L. C. Devinney, S. A. Star, and R. M. Williams Jr. 1949. *The American Soldier: Adjustment during Army Life. (Studies in Social Psychology in World War II), Vol. 1.* xii, 599. Oxford, UK: Princeton Univ. Press.

Strogatz, S. H. 1988. "Love Affairs and Differential Equations." *Mathematics Magazine* 61 (1): 35–35. https://doi.org/10.1080/0025570X.1988.11977342.

———. 2018. *Nonlinear Dynamics and Chaos with Student Solutions Manual: With Applications to Physics, Biology, Chemistry, and Engineering, Second Edition.* Boca Raton, FL: CRC Press.

Surowiecki, J. 2005. *The Wisdom of Crowds.* New York, NY: Knopf Doubleday Publishing Group.

Sussmann, H. J., and R. S. Zahler. 1978. "Catastrophe Theory as Applied to the Social and Biological Sciences: A Critique." *Synthese* 37 (2): 117–216. https://doi.org/10.1007/BF00869575.

Sutton, R. S., and A. G. Barto. 2018. *Reinforcement Learning, Second Edition: An Introduction.* Cambridge, MA: MIT Press.

Sznajd-Weron, Katarzyna, J. Sznajd, and T. Weron. 2021. "A Review on the Sznajd Model — 20 Years After." *Physica A: Statistical Mechanics and its Applications* 565:125537. https://doi.org/10.1016/j.physa.2020.125537.

Szostak, N., S. Wasik, and J. Blazewicz. 2016. "Hypercycle." Edited by S. Wodak. *PLOS Computational Biology* 12 (4): e1004853. https://doi.org/10.1371/journal.pcbi.1004853.

Tesfatsion, L. 2002. "Agent-Based Computational Economics: Growing Economies From the Bottom Up." *Artificial Life* 8 (1): 55–82. https://doi.org/10.1162/106454602753694765.

Tesser, A. 1978. "Self-Generated Attitude." In *Advances in Experimental Social Psychology,* edited by L. Berkowitz, 11:289–338. Academic Press. https://doi.org/10.1016/S0065-2601(08)60010-6.

Thelen, E. 1995. "Motor Development: A New Synthesis." *American Psychologist* 50:79–95. https://doi.org/10.1037/0003-066X.50.2.79.

Thelen, E., and L. B. Smith. 1994. *A Dynamic Systems Approach to the Development of Cognition and Action.* Cambridge, MA: MIT Press.

Thom, R. 1977. "Structural Stability, Catastrophe Theory, and Applied Mathematics." *SIAM Review* 19 (2): 189–201. https://doi.org/10.1137/1019036.

———. 2018. *Structural Stability And Morphogenesis.* Cambridge, MA: CRC Press.

Thompson, M. P., H. M. Farrell, and R. Greenberg. 1969. "Alpha S1-Casein A (Bos Taurus): A Probable Sequential Deletion of Eight Amino Acid Residues and Its Effect on Physical Properties." *Comparative Biochemistry and Physiology* 28 (1): 471–475. https://doi.org/10.1016/0010-406x(69)91364-4.

Thurner, S. 2017. *43 Visions For Complexity.* Singapore: World Scientific.

Thurner, S., P. Klimek, and R. Hanel. 2018. *Introduction to the Theory of Complex Systems.* Oxford University Press. https://doi.org/10.1093/oso/9780198821939.001.0001.

Tilman, D., S. S. Kilham, and P. Kilham. 1982. "Phytoplankton Community Ecology: The Role of Limiting Nutrients." *Annual Review of Ecology and Systematics* 13 (1): 349–372. https://doi.org/10.1146/annurev.es.13.110182.002025.

Toker, D., F. T. Sommer, and M. D'Esposito. 2020. "A Simple Method for Detecting Chaos in Nature." *Communications Biology* 3 (1): 1–13. https://doi.org/10.1038/s42003-019-0715-9.

Tononi, G., O. Sporns, and G. M. Edelman. 1994. "A Measure for Brain Complexity: Relating Functional Segregation and Integration in the Nervous System." *Proceedings of the National Academy of Sciences of the United States of America* 91 (11): 5033–5037. https://doi.org/10.1073/pnas.91.11.5033.

Treiber, M., A. Hennecke, and D. Helbing. 2000. "Congested Traffic States in Empirical Observations and Microscopic Simulations." *Physical Review E* 62 (2): 1805–1824. https://doi.org/10.1103/PhysRevE.62.1805.

Treur, J. 2019. *Network-Oriented Modeling for Adaptive Networks: Designing Higher-Order Adaptive Biological, Mental and Social Network Models.* 1st ed. 2020 edition. Springer.

Tschacher, W., and H. Haken. 2019. *The Process of Psychotherapy: Causation and Chance.* Springer.

———. 2023. "A Complexity Science Account of Humor." *Entropy* 25 (2): 341. https://doi.org/10.3390/e25020341.

Tuerlinckx, F., E. Maris, R. Ratcliff, and P. De Boeck. 2001. "A Comparison of Four Methods for Simulating the Diffusion Process." *Behavior Research Methods, Instruments, & Computers* 33 (4): 443–456. https://doi.org/10.3758/BF03195402.

Turner-Zwinkels, F. M., and M. J. Brandt. 2022. "Belief System Networks Can Be Used to Predict Where to Expect Dynamic Constraint." *Journal of Experimental Social Psychology* 100:104279. https://doi.org/10.1016/j.jesp.2021.104279.

Tushman, M. L., and P. W. Anderson. 1986. "Technological Discontinuities and Organizational Environments." *Administrative Science Quarterly* 31 (3): 439–465. https://doi.org/10.2307/2392832.

Tyutyunov, Y. V., and L. I. Titova. 2020. "From Lotka–Volterra to Arditi–Ginzburg: 90 Years of Evolving Trophic Functions." *Biology Bulletin Reviews* 10 (3): 167–185. https://doi.org/10.1134/S207908642003007X.

Uher, J. 2021. "Psychometrics Is Not Measurement: Unraveling a Fundamental Misconception in Quantitative Psychology and the Complex Network of Its Underlying Fallacies." *Journal of Theoretical and Philosophical Psychology* 41:58–84. https://doi.org/10.1037/teo0000176.

Vallacher, R. R., and A. Nowak, eds. 1994. *Dynamical Systems in Social Psychology.* xx, 305. San Diego, CA.

Vallacher, R. R., S. J. Read, and A. Nowak, eds. 2017. *Computational Social Psychology.* 1st edition. New York, NY; London, UK.

van Bork, R., S. Epskamp, M. Rhemtulla, D. Borsboom, and H. L. J. van der Maas. 2017. "What Is the P-Factor of Psychopathology? Some Risks of General Factor Modeling." *Theory & Psychology* 27 (6): 759–773. https://doi.org/10.1177/0959354317737185.

van de Leemput, I. A., M. Wichers, A. O. J. Cramer, D. Borsboom, F. Tuerlinckx, P. Kuppens, E. H. van Nes, et al. 2014. "Critical Slowing down as Early Warning for the Onset and Termination of Depression." *Proceedings of the National Academy of Sciences* 111 (1): 87–92. https://doi.org/10.1073/pnas.1312114110.

van den Burg, G. J. J., and C. K. I. Williams. 2022. *An Evaluation of Change Point Detection Algorithms.* https://doi.org/10.48550/arXiv.2003.06222.

van den Ende, M. W. J., S. Epskamp, M. H. Lees, H. L. J. van der Maas, R. W. Wiers, and P. M. A. Sloot. 2022. "A Review of Mathematical Modeling of Addiction Regarding Both (Neuro-) Psychological Processes and the Social Contagion Perspectives." *Addictive Behaviors* 127:107201. https://doi.org/10.1016/j.addbeh.2021.107201.

van der Maas, H. L. J. 1995. "Beyond the Metaphor?" *Cognitive Development* 10:621–642.

———. 2022. "Spiegeloog - A Dialogue on Consciousness." *Spiegeloog,* https://www.spiegeloog.amsterdam/a-dialogue-on-consciousness/.

van der Maas, H. L. J., J. Dalege, and L. J. Waldorp. 2020. "The Polarization within and across Individuals: The Hierarchical Ising Opinion Model." *Journal of Complex Networks* 8 (2): cnaa010. https://doi.org/10.1093/comnet/cnaa010.

van der Maas, H. L. J., C. V. Dolan, R. P. P. P. Grasman, J. M. Wicherts, H. M. Huizenga, and M. E. J. Raijmakers. 2006. "A Dynamical Model of General Intelligence: The Positive Manifold of Intelligence by Mutualism." *Psychological Review* 113 (4): 842–861. https://doi.org/10/c3jm44.

van der Maas, H. L. J., K.-J. Kan, and D. Borsboom. 2014. "Intelligence Is What the Intelligence Test Measures. Seriously." *Journal of Intelligence* 2 (1): 12–15. https://doi.org/10.3390/jintelligence2010012.

van der Maas, H. L. J., K.-J. Kan, M. Marsman, and C. E. Stevenson. 2017. "Network Models for Cognitive Development and Intelligence." *Journal of Intelligence* 5 (2): 16. https://doi.org/10.3390/jintelligence5020016.

van der Maas, H. L. J., R. Kolstein, and J. van der Pligt. 2003. "Sudden Transitions in Attitudes." *Sociological Methods & Research* 32 (2): 125–152. https://doi.org/10.1177/0049124103253773.

van der Maas, H. L. J., and P. C. M. Molenaar. 1992. "Stagewise Cognitive Development: An Application of Catastrophe Theory." *Psychological Review* 99 (3): 395–417. https://doi.org/10.1037/0033-295X.99.3.395.

van der Maas, H. L. J., L. Snoek, and C. E. Stevenson. 2021. "How Much Intelligence Is There in Artificial Intelligence? A 2020 Update." *Intelligence* 87:101548. https://doi.org/10.1016/j.intell.2021.101548.

van der Maas, H. L. J., P. F. M. J. Verschure, and P. C. M. Molenaar. 1990. "A Note on Chaotic Behavior in Simple Neural Networks." *Neural Networks* 3 (1): 119–122. https://doi.org/10.1016/0893-6080(90)90050-U.

van Dongen, N., R. van Bork, A. Finnemann, H. L. J. van der Maas, D. J. Robinaugh, J. Haslbeck, J. de Ron, J. Sprenger, and D. Borsboom. 2022. "Productive Explanation: A Framework for Evaluating Explanations in Psychological Scienc." *Psychologial Review,* https://doi.org/10.31234/osf.io/qd69g.

van Geert, P. 1991. "A Dynamic Systems Model of Cognitive and Language Growth." *Psychological Review* 98:3–53. https://doi.org/10.1037/0033-295X.98.1.3.

———. 1998. "A Dynamic Systems Model of Basic Developmental Mechanisms: Piaget, Vygotsky, and Beyond." *Psychological Review* 105:634–677. https://doi.org/10.1037/0033-295X.105.4.634-677.

van Ooyen, A., and M. Butz-Ostendorf. 2017. *The Rewiring Brain: A Computational Approach to Structural Plasticity in the Adult Brain.* London, UK: Academic Press.

van Overwalle, F., and F. Siebler. 2005. "A Connectionist Model of Attitude Formation and Change." *Personality and Social Psychology Review: An Official Journal of the Society for Personality and Social Psychology, Inc* 9 (3): 231–274. https://doi.org/10.1207/s15327957pspr0903_3.

van Rooij, I., and M. Blokpoel. 2020. "Formalizing Verbal Theories." *Social Psychology* 51 (5): 285–298. https://doi.org/10.1027/1864-9335/a000428.

Vangeli, E., J. Stapleton, E. S. Smit, R. Borland, and R. West. 2011. "Predictors of Attempts to Stop Smoking and Their Success in Adult General Population Samples: A Systematic Review." *Addiction* 106 (12): 2110–2121. https://doi.org/10.1111/j.1360-0443.2011.03565.x.

Varela, F. J., E. Thompson, and E. Rosch. 2017. *The Embodied Mind, Revised Edition: Cognitive Science and Human Experience.* MIT Press.

Verdonck, S., and F. Tuerlinckx. 2014. "The Ising Decision Maker: A Binary Stochastic Network for Choice Response Time." *Psychological Review* 121:422–462. https://doi.org/10.1037/a0037012.

Visser, I., and M. Speekenbrink. 2022. *Mixture and Hidden Markov Models with R.* Cham, Switzerland: Springer International Publishing.

Voelkle, M. C., J. H. L. Oud, E. Davidov, and P. Schmidt. 2012. "An SEM Approach to Continuous Time Modeling of Panel Data: Relating Authoritarianism and Anomia." *Psychological Methods* 17:176–192. https://doi.org/10.1037/a0027543.

Volberda, H. W., and A. Y. Lewin. 2003. "Co-Evolutionary Dynamics Within and Between Firms: From Evolution to Co-evolution." *Journal of Management Studies* 40 (8): 2111–2136. https://doi.org/10.1046/j.1467-6486.2003.00414.x.

Von Bertalanffy, L. 1969. *General System Theory: Foundations, Development, Applications.* Revised edition. New York, NY: George Braziller Inc.

von der Heydt, A., M. Dekker, and H. Dijkstra. 2019. "Cascading Transitions in the Climate System." *Earth System Dynamics* 9 (4): 10435. https://doi.org/10.5194/esd-9-1243-2018.

Wagemans, J., J. H. Elder, M. Kubovy, S. E. Palmer, M. A. Peterson, M. Singh, and R. von der Heydt. 2012. "A Century of Gestalt Psychology in Visual Perception: I. Perceptual Grouping and Figure–Ground Organization." *Psychological Bulletin* 138:1172–1217. https://doi.org/10.1037/a0029333.

Wagenmakers, E.-J., P. C. M. Molenaar, R. P. P. P. Grasman, P. A. I. Hartelman, and H. L. J. van der Maas. 2005. "Transformation Invariant Stochastic Catastrophe Theory." *Physica D: Nonlinear Phenomena* 211 (3): 263–276. https://doi.org/10.1016/j.physd.2005.08.014.

Waldorp, L. J., and M. Marsman. 2022. "Relations between Networks, Regression, Partial Correlation, and the Latent Variable Model." *Multivariate Behavioral Research* 57 (6): 994–1006. https://doi.org/10.1080/00273171.2021.1938959.

Weaver, W. 1948. "Science and Complexity." *American Scientist* 36 (4): 536–544. https://www.jstor.org/stable/27826254.

Weisbuch, G., G. Deffuant, F. Amblard, and J.-P. Nadal. 2002. "Meet, Discuss, and Segregate!" *Complexity* 7 (3): 55–63. https://doi.org/10.1002/cplx.10031.

Wheeler, J. A. 1994. *At Home in the Universe.* New York, NY: American Institute of Physics.

Wichers, M., P. C. Groot, and ESM Group Psychosystems. 2016. "Critical Slowing Down as a Personalized Early Warning Signal for Depression." *Psychotherapy and Psychosomatics* 85 (2): 114–116. https://doi.org/10.1159/000441458.

Wiener, N. 2019. *Cybernetics or Control and Communication in the Animal and the Machine, Reissue of the 1961 Second Edition.* Cambridge, MA: MIT Press.

Wilensky, U., and W. Rand. 2015. *An Introduction to Agent-Based Modeling: Modeling Natural, Social, and Engineered Complex Systems with NetLogo.* Cambridge, MA: The MIT Press.

Wimsatt, William C. 1994. "The Ontology of Complex Systems: Levels of Organization, Perspectives, and Causal Thickets." *Canadian Journal of Philosophy Supplementary Volume* 20:207–274. https://doi.org/10.1080/00455091.1994.10717400.

Witkiewitz, K., H. L. J. van der Maas, M. R. Hufford, and G. A. Marlatt. 2007. "Nonnormality and Divergence in Posttreatment Alcohol Use: Reexamining the Project MATCH Data "Another Way."" *Journal of Abnormal Psychology* 116:378–394. https://doi.org/10.1037/0021-843X.116.2.378.

Wittenborn, A. K., H. Rahmandad, J. Rick, and N. Hosseinichimeh. 2016. "Depression as a Systemic Syndrome: Mapping the Feedback Loops of Major Depressive Disorder." *Psychological Medicine* 46 (3): 551–562. https://doi.org/10.1017/S0033291715002044.

Xue, J., and B. Shen. 2020. "A Novel Swarm Intelligence Optimization Approach: Sparrow Search Algorithm." *Systems Science & Control Engineering* 8 (1): 22–34. https://doi.org/10.1080/21642583.2019.1708830.

Young, G. 2015. "Causality in Psychiatry: A Hybrid Symptom Network Construct Model." *Frontiers in Psychiatry* 6. https://doi.org/10.3389/fpsyt.2015.00164.

Zahler, R. S., and H. J. Sussmann. 1977. "Claims and Accomplishments of Applied Catastrophe Theory." *Nature* 269:759–763. https://doi.org/10.1038/269759a0.

Zanin, M. 2022. "Can Deep Learning distinguish chaos from noise? Numerical experiments and general considerations." *Communications in Nonlinear Science and Numerical Simulation* 114:106708. https://doi.org/10.1016/j.cnsns.2022.106708.

Zeeman, E. C. 1976. "Catastrophe Theory." *Scientific American* 234 (4): 65–83.

Zhang, J., and Y. Zhao. 2018. "The Robust Consensus of a Noisy Deffuant-Weisbuch Model." *Mathematical Problems in Engineering* 2018:e1065451. https://doi.org/10.1155/2018/1065451.

Zwicker, M. V., H. U. Nohlen, J. Dalege, G.-J. M. Gruter, and F. van Harreveld. 2020. "Applying an Attitude Network Approach to Consumer Behaviour towards Plastic." *Journal of Environmental Psychology* 69:101433. https://doi.org/10.1016/j.jenvp.2020.101433.

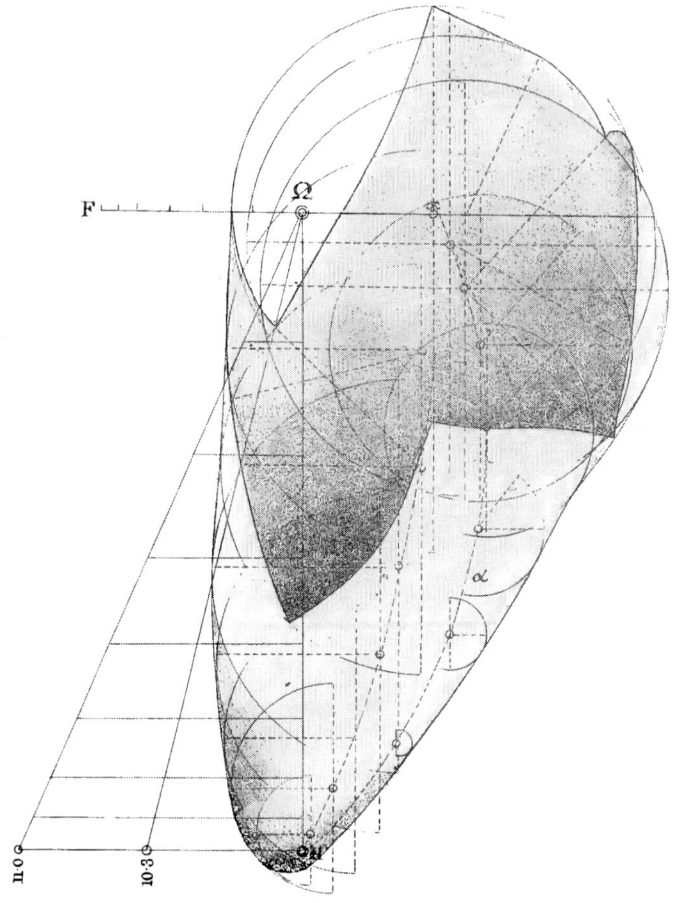

F
Ω
α
11·0
10·3

THE IMAGES IN THIS VOLUME

The images in this volume were originally published in Louisa S. Cook's now-public-domain book, *Geometrical Psychology, or the Science of Representation: An Abstract of the Theories and Diagrams of B. W. Betts* (1887).

Born in 1832, Betts initially worked as an architect in England before heading abroad. While working as a "trigonometrical computer" for the civil service in Auckland, New Zealand, he became preoccupied with notions of truth and form. Betts drew inspiration from Johann Gottlieb Fichte's suggestion, in the *Grundlage der gesammten Wissenschaftslehre* (1794/5), that geometric concepts such as the line or circle could correspond to modes of consciousness, and developed the plane forms reproduced here. Betts shared his diagrams and corresponded at length with the self-taught mathematician Mary Everest Boole, wife of the late George Boole.

Cook describes his work as follows (pp. 9–11):

> *He contemplates Man, not from the physical, but from the metaphysical point of view; thus the evolution of Man is for him the evolution of human consciousness. He attempts to represent the successive stages of this evolution by means of symbolical methmatical forms. These forms represent the course of development of human consciousness from the animal basis, the pure sense-consciousness, to the spiritual or divine consciousness; both which extremes are not man—the one underlying, the other transcending the limits of human evolution.*
>
> *[...]*
>
> *The symbolic forms which Mr. Betts has evolved through his system of Representation resemble, when developed in two dimensions, conventionalised but very scientifically and beautifully conventionalised leaf-outlines. When in more than two dimensions they approximate to the forms of flowers and crystals.*

A scan of Cook's abstract is available to view and read courtesy of the *Internet Archive*: https://archive.org/details/geometricalpsychoocook.

AUTHOR

HAN L. J. VAN DER MAAS is Professor of Psychological Methods in the Department of Psychology at the University of Amsterdam and Distinguished Research Professor of Complex Systems in the Social and Behavioral Sciences. He is also an External Faculty Member of the Santa Fe Institute and a Principal Investigator and Board Member of the Institute of Advanced Study at the UvA.

Van der Maas received his PhD in developmental psychology in 1993, focusing on phase transitions in cognitive development. He joined the faculty of the University of Amsterdam, becoming full professor in 2003 and chair of the Psychological Methods Group in 2005. He has held leadership positions as Director of the Graduate School and Director of Research.

In 2009 he founded Oefenweb.nl, a spin-off company that developed a game-based adaptive child monitoring system, now part of Prowise.com. His research focuses on formalizing and testing psychological theories of cognition, expertise, development, attitudes, and intelligence. In 2022, he received an ERC advanced grant to study cascading transitions in psychosocial systems.

SFI PRESS

THE SANTA FE INSTITUTE PRESS

The SFI Press endeavors to communicate the best of complexity science and to capture a sense of the diversity, range, breadth, excitement, and ambition of research at the Santa Fe Institute. To provide a distillation of discussions, debates, and meetings across a range of influential and nascent topics.

To change the way we think.

SEMINAR SERIES

New findings emerging from the Institute's ongoing working groups and research projects, for an audience of interdisciplinary scholars and practitioners.

ARCHIVE SERIES

Fresh editions of classic texts from the complexity canon, spanning the Institute's thirty years advancing the field.

COMPASS SERIES

Provocative, exploratory volumes aiming to build complexity literacy in the humanities, industry, and the curious public.

SCHOLARS SERIES

Affordable and accessible textbooks disseminating the latest findings in the complex systems science world.

Also from SFI Press

Agent-Based Modeling for Archaeology: Simulating the Complexity of Societies
Iza Romanowska, Colin D. Wren, and Stefani A. Crabtree

Worlds Hidden in Plain Sight: The Evolving Idea of Complexity at the Santa Fe Institute, 1984-2019
David C. Krakauer, ed.

For additional titles, inquiries, or news about the Press, visit us at
WWW.SFIPRESS.ORG.

ABOUT THE SANTA FE INSTITUTE

The Santa Fe Institute is the world headquarters for complexity science, operated as an independent, nonprofit research and education center located in Santa Fe, New Mexico. Our researchers endeavor to understand and unify the underlying, shared patterns in complex physical, biological, social, cultural, technological, and even possible astrobiological worlds. Our global research network of scholars spans borders, departments, and disciplines, bringing together curious minds steeped in rigorous logical, mathematical, and computational reasoning. As we reveal the unseen mechanisms and processes that shape these evolving worlds, we seek to use this understanding to promote the well-being of humankind and of life on Earth.

COLOPHON

The body copy for this book was set in EB Garamond, a typeface designed by Georg Duffner after the Ebenolff-Berner type specimen of 1592. Headings are in Kurier, created by Janusz M. Nowacki, based on typefaces by the Polish typographer Małgorzata Budyta. For footnotes and captions, we have used CMU Bright, a sans serif variant of Computer Modern, created by Donald Knuth for use in TeX, the typesetting program he developed in 1978.

The SFI Press complexity glyphs used throughout this book were designed by Brian Crandall Williams.

SANTA FE INSTITUTE

COMPLEXITY GLYPHS

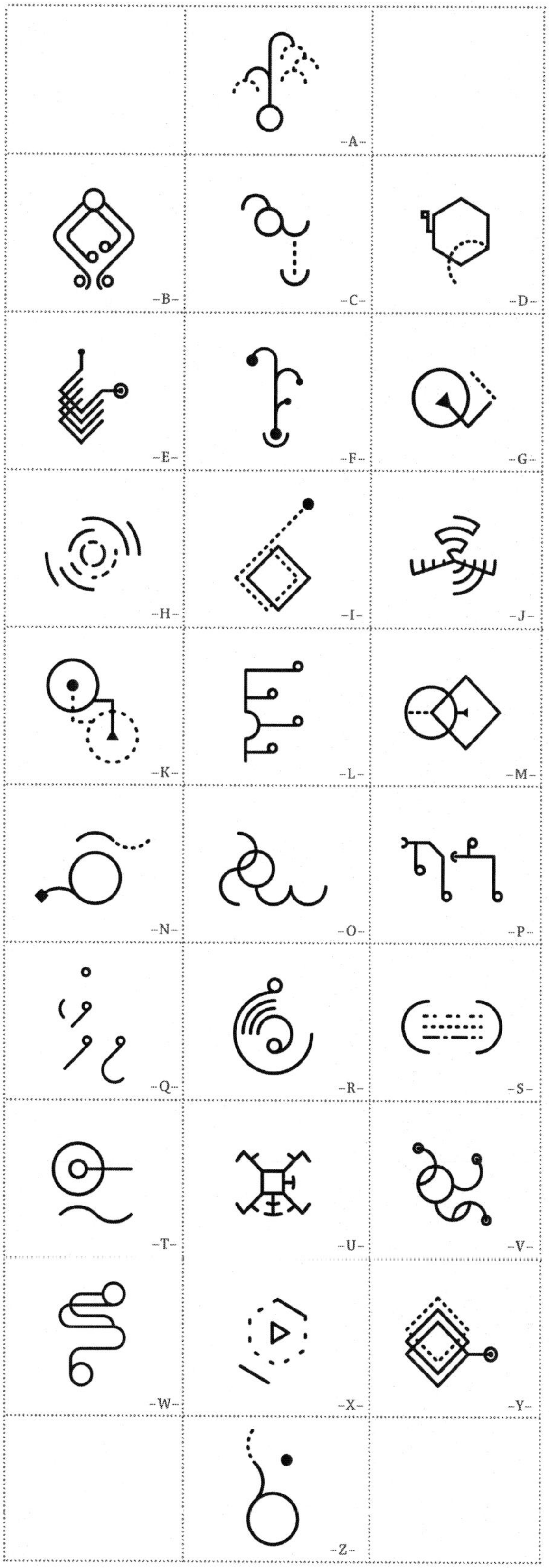

SFI PRESS

SCHOLARS SERIES

Made in United States
North Haven, CT
10 November 2026

22097466R00180